BABAO 2004

Proceedings of the 6th Annual Conference of the
British Association for Biological Anthropology and
Osteoarchaeology, University of Bristol

Edited by

Kate A. Robson Brown
Alice M. Roberts

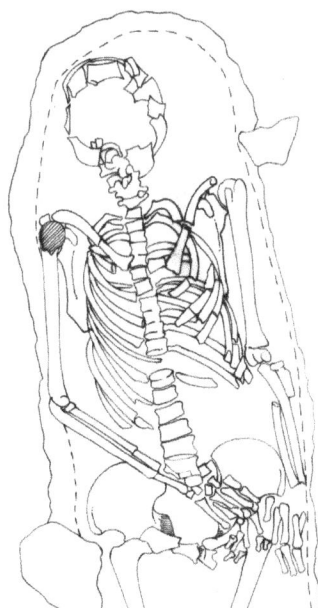

BAR International Series 1623
2007

Published in 2016 by
BAR Publishing, Oxford

BAR International Series 1623

BABAO 2004

ISBN 978 1 4073 0035 1

BAR Publishing is the trading name of British Archaeological Reports (Oxford) Ltd.
British Archaeological Reports was first incorporated in 1974 to publish the BAR
Series, International and British. In 1992 Hadrian Books Ltd became part of the BAR
group. This volume was originally published by Archaeopress in conjunction with
British Archaeological Reports (Oxford) Ltd / Hadrian Books Ltd, the Series principal
publisher, in 2007. This present volume is published by BAR Publishing, 2016.

Printed in England

BAR
PUBLISHING

BAR titles are available from:

	BAR Publishing
	122 Banbury Rd, Oxford, OX2 7BP, UK
EMAIL	info@barpublishing.com
PHONE	+44 (0)1865 310431
FAX	+44 (0)1865 316916
	www.barpublishing.com

CONTENTS

PREFACE

On the morning of 10th September 2004, the University of Bristol opened its doors to BABAO for the first time. Delegates from across the country, and representing many sub-disciplines of our broad field, were welcomed into Southwell Street conference centre for what was to become a very dynamic and open meeting, full of goodwill and collaboration.

The University has long been associated with the developing disciplines of biological anthropology and osteoarchaeology. The late and much missed Dr Juliet Rogers, in the Department of Rheumatology, was a pioneering researcher in palaeopathology and mentor for young osteoarchaeologists, and her legacy lives on in the continuing strong collaborative links between the Department of Archaeology and Anthropology, the Bristol Royal Infirmary, the Department of Anatomy, and archaeological units in the South West region. In a similar way Dr Jonathan Musgrave, during his long career in the Medical School, has created a hub for evolutionary and forensic anthropology that grows from strength to strength. It is a measure of the University's commitment to these and related fields that over the past decade new appointments and facilities have fostered the growth of research groups that continue to bridge the artificial divides between Faculties and institutions. In 2003 we were eager to take on the challenge of hosting BABAO 2004, aiming to organise a broad and inclusive meeting to reflect the particularly interdisciplinary character of biological anthropology and osteoarchaeology at Bristol.

Over the two days delegates were treated to 37 podium and 9 poster presentations, plus a fascinating demonstration of butchery practices in the dissection room. Professor Fred Spoor introduced the meeting in great style, with his paper on the new hominin fossils recently discovered in the Koobi Fora Formation. Following on from this, the rest of the presentations were divided into 6 themed sessions. The first two focussed on Evolutionary Anthropology, covering the broad themes of the evolution and development of the skull, and the comparative primate context. In the session on human behavioural ecology, papers on parental investment, reproductive strategies and child health were presented. Osteoarchaeology was central to three sessions; the first of these focussed on palaeodemography, with particular emphasis on biomolecular techniques. The second explored human skeletal peri and post mortem modification, and included papers advocating an approach integrating zooarchaeology, taphonomy, and pathology. The third osteoarchaeology session related to burial practices, and in particular the issues of identity and preservation. The last session, on palaeopathology, was dedicated to the memory of Dr Juliet Rogers.

This volume presents selected papers from across the sessions. The interdisciplinary character of the meeting, and the healthy state of British biological anthropology as a whole, is reflected well in them all.

K.A. ROBSON BROWN
A.M. ROBERTS
Bristol, April 2006

LIST OF CONTRIBUTORS

C. Buckley
Archaeology Research Group
School of Humanities
University of Southampton
SO17 1BF

A.T. Chamberlain
Department of Archaeology
University of Sheffield
Northgate House
West Street
Sheffield
S1 4ET

C. Chapman
Department of Anthropology
University of Kent
Canterbury
Kent
CT2 7NR

C.A. Chenery
NIGL
British Geological Survey
Kingsley Durnham Centre
Keyworth
Nottingham
NG12 5GG

M. Clegg
Centre for the Archaeology of Human Origin
University of Southampton

A. Gray Jones
Spitalfields Project
Museum of London Specialist Services
46 Eagle Wharf Road
London
N1 7ED

S.E. Johns
Department of Anthropology
University of Kent
Canterbury
Kent
CT2 7NR

S.S. Legge
Department of Anthropology
University of Kent
Canterbury
Kent
CT2 7NS

P.M. Macpherson
Department of Archaeology
University of Sheffield
Northgate House
West Street
Sheffield
S1 4ET

X.D.G. Mallett
Human Identification
Sheffield University
School of Medicine
Sheffield
S10 2RX

A.M. Roberts
Department of Anatomy
Southwell Street,
Bristol
BS2 8EJ

K. Robson Brown
Department of Archaeology and Anthropology
43 Woodland Road
Bristol
BS8 1UU

K. Seetah
Grahame Clark Zooarchaeology Laboratory
Department of Archaeology
Cambridge University
Downing Street
Cambridge
CB2 3DZ

D. Walker
Spitalfields Project
Museum of London Specialist Services
46 Eagle Wharf Road
London
N1 7ED

S. R. Zakrzewski
Department of Archaeology
Highfield
Southampton
SO17 1BF

CLIMATIC INFLUENCES ON CRANIOFACIAL VARIABILITY IN MODERN HUMANS

X.D.G. Mallett

Abstract

There is considerable evidence that craniofacial morphology is responsive to the climatic gradient of the immediate environment in which individuals live. This study explores the process of differentiation among modern Amerindians, to determine if there is osteometric evidence for morphological characteristics that correspond with long-term climatic adaptation. Two predictive hypothetical models were developed; the Climatic Adaptation and Genetic Models, tested using craniometric data gathered from five distinct Native American foraging groups. The groups consisted of the cold-climate Greenland Eskimo, Newfoundland Inuit, and Argentinean Indians, and the temperate-climate Californian and New Mexican Indians. Analysis of Variance (ANOVA) and Canonical Variates Analyses (CVA) were used to investigate divergence between 20 variables to analyse size variation and seven indices assessing shape differences. The ANOVA results demonstrated that there is morphological convergence between the cold-climate groups in several variables. Specifically, maximum cranial length and naso-dacryon subtense had the most discriminant power. Further, the results generally support the interpretation that face size is linked to cold-climatic adaptation. The CVA results for shape demonstrate that only orbital index is of discriminatory value when differentiating between warm- and cold-climate groups. The results for the remainder of the craniofacial skeleton did not fit either model, and the balance of facial shape in these populations appears to be the result of an extraneous influence not tested for in this study. Overall, the results suggest that climate is more influential in determining craniofacial size in cold-climates than was previously predicted. This signifies that there is potential for climate-led morphological analyses to be used to infer processes of adaptation to environment.

Key words: Amerindian; ANOVA; craniofacial variability; cold-adaptation; CVA.

Introduction

For over a century it has been understood that humans and other mammals adapt to external pressures placed upon them, both immediately to the environment surrounding individuals and on a population level, with plastic changes taking place within the skeleton. As a species, humans show incredible morphological diversity (Anton 1989; Biggers *et al.* 1958; Howells 1960a, 1960b; Ogle 1934; Summer 1909). Much of the biological variation among populations involves modest degrees of difference in the frequency of shared traits, reflecting both hereditary factors and the influence of natural and social environments (Walker 1996). Physical

differentiation of modern human groups is therefore not strictly random. The influences of climate, altitude, and geographic location all contribute to the varying morphology of populations, and previous research suggests that several aspects of modern human craniofacial variability are the result of long-term adaptation to climate (Churchill 1998; Guglielmino-Matessi *et al.* 1979; Hernández *et al.* 1997a; Hylander 1977; Lahr 1995; Lalueza *et al.* 1997; Lieberman *et al.* 2000).

When assessing adaptation to environment, geographical isolation is often a requisite of the populations selected, as it has been suggested that the geographical isolation of small local populations exposed to climatic extremes can accelerate the rate of morphological changes, therefore accentuating the development of divisions among population groups (Coon 1971; Lalueza *et al.* 1997). Consequently, the study of isolated groups facilitates the assessment of human plasticity and adaptability, as well as degrees of morphological variability.

For many years authors have sought to find relationships between human adaptation and the climate of the environment which they have inhabited (Foley 1994; Vrba 1980). Specifically, craniofacial adaptation has been the subject of much interest, with many studies concentrating on analysing the "cold-adapted" face, as humans have adapted to environmental extremes (Blumenfeld 2002; Churchill 1998; Franciscus 2003; Friess 2002; Hernández *et al.* 1997a; Steegman 1965, 1970a, 1972, 1975; Steegman & Platner 1968).

As a result of this research, the following morphological features of the skull have been hypothesised to correspond with adaptation to cold climates: mid-facial prognathism, malar size and projection, nasal dimensions, a long head and face with transverse flatness, and maxillary sinus (Hanihara 1979, 1993a, 1993b, 1996; Hanihara 2000; Howells 1973; Lahr 1995; Lahr & Wright 1996; Lalueza *et al.* 1997; Steegman *et al.* 2002; Wolpoff 1968). The correlation of these features have been described as a "flat face" (Hanihara 2000; Hernández *et al.* 1997a), and is commonly associated with people from cold climates (Hanihara 2000; Wolpoff 1968). This morphology is very distinctive, with malar enlargement and protrusion dominating facial form (Churchill 1998; Steegman 1972). However, explicitly isolating the morphological features that are expected to vary with climate is difficult, as much variation occurs due to cultural and genetic influence. Steegman (1970a) recognises only one feature as specifically reflecting adaptation to cold environments; a narrow and vertically high nose. This may be associated with climatic factors due to the nose's functional relationship with thermoregulation (Franciscus 2003; Hernández *et al.* 1997a). However, the Fuèguians and Eskimos (two populations thought to be adapted to cold) differ significantly in most nasal dimensions, including nasal index, minimum breadth, naso-frontal breadth, and inter-orbital breadth. As Fuèguians and Eskimos are not closely associated either genetically or geographically, a possible explanation of similar nasal height measurements is that the nasal morphology in both groups is the result of adaptive pressures related to cold stress (Lalueza *et al.* 1997).

The aim of this study is to investigate whether the morphological features of the skull which have been hypothesized to correspond with cold-climate adaptation, a) correspond and co-vary with one another, and b) correspond with climatic gradients.

Waves of American Migration
Research relating to the peopling of the New World suggests that the most probable model to explain the extant genetic diversity is a demic diffusion from north to south, starting in Siberia and finishing in the extreme south of South America (Neves & Pucciarelli 1991). This theory is based on the concept of discrete migrations, in a number of waves, into the Americas from the Old World (Powell & Neves 1999). A number of hypotheses are currently being debated in the literature relating to the timing of the American waves of migration. However, the most commonly accepted model suggests that the continent was occupied by more than one group originally, before and during the last glacial (Lahr 1995).

Further to this, the Three-Migration Model, developed from this migrationary hypothesis, suggests that tripartite waves of people entered America and populated it in three distinct periods. The first wave likely entered the New World around 10-15,000 B.P. (Adams *et al.* 2001; González-José *et al.* 2001; Hernández *et al.* 1997a; Lalueza *et al.* 1997; Steele *et al.* 1998), and included the people who settled in the southern-most tips of Argentina and Chile. The second wave is thought to have occurred at around 9,000 B.P, consisting of the people who came to settle in mid-America (Greenberg *et al.* 1986). The Eskimo represent a late and final wave of migration into America at approximately ≤5,000 B. P. (Greenberg *et al.* 1986). The differences in models, as well as the timing of the waves of migration and adaptive radiations that are thought to have occurred, have vast implications when attempting to investigate human evolution.

Alternate Hypotheses
To answer the questions raised in this paper two hypotheses were developed, and the selection of the five skeletal samples made with these in mind. The hypotheses represent trends in facial anatomy, predicted under models of climatic or genetic influences on morphological variability; these have been refined into two alternate prognostic models, developed to further our understanding of human adaptability, based on the assumption of regional continuity in Amerindian

populations as all of the population groups selected are of Amerindian descent.

The Climatic Adaptation Model
The Climatic Adaptation Model (CAM) is founded upon a theory developed from previous studies of cold adaptation (Howells 1973, 1989; Lahr 1995, 1996; Lalueza *et al.* 1997; Steegman 1970a, 1975; Steegman *et al.* 2002) and interpretations of Neanderthal morphology (Blumenfeld 2002; Churchill 1998). It expresses the results of convergent adaptation, and is based upon the premise that climate will be largely responsible for the morphological features present. If climatic differences between geographical regions have influenced the craniofacial anatomy of the selected populations, we would expect morphological convergence between the populations from cold environments. Evidence of this convergence will take the form of a distinct suite of features apparent in all three cold-climate groups, the origin of which could not be genetic as the populations were selected for their relative genetic isolation. In the statistical analyses, the populations will cluster by temperature of the region from which they originate. The results will form two clusters: the two Eskimo groups (Angmaksaslik & Labradorean Inuit) and the Argentinean (Fuèguian) sample will form cluster 1, and the Californian (Chumash) and New Mexican (Ketchipauan) sets cluster 2, as shown in Figure 1. Consequently, the model predicts that the group from the most northern region, the Greenland Eskimo, will be most similar to the most southern group, the Argentinean set, in morphological trait expression. Similarly, the Californian and New Mexican Groups will show a large divergence in facial characteristics from the three cold-climate groups.

As the Greenland (Angmaksaslik) and Argentinean (Fuèguian) samples are located at the extreme northern and southern tips of America and Greenland, it is reasonable to assume that a number of features must have occurred as consequences of ontogenetic programmes that were geared to produce cold-adapted body forms, as the high level of human adaptive plasticity bears evidence (Anton 1989; Howells 1960a, 1960b): This is because no genetic influence could have affected the results as the sample from Argentina (Tierra del Fuègo) is now generally accepted to have entered the New World around 10-15,000 years ago (Colvin 2002; González-José *et al.* 2001; Hernández *et al.* 1997a; Lalueza *et al.* 1997) and been isolated for approximately 8,000 B.P. years, until European contact (Fox *et al.* 1996). Conversely, the Greenland Eskimo (Angmaksaslik) are a late wave of migration into America at approximately 5,000 B.P. (Greenberg *et al.* 1986).

Attributes expected to feature most prominently in distinguishing populations from cold environments are: a long head and face with pronounced transverse flatness and mid-facial prognathism produced by long upper facial height, basion-prosthion and basion-bregmatic lengths, and deep cheek height. The narrow head is formed by narrow maximum frontal breadth, fronto-malar breadth, minimum frontal breadth, and maximum cranial breadth. Mid-facial prognathism is the result of broad bi-maxillary, bi-jugal, bi-zygomatic, and bi-condylar breadths. The long slim nose is produced by long nasal height and nasal breadth, naso-dacryon subtense and narrow inter-orbital breadth. Furthermore, malar enlargement and protrusion will be present, dominating facial form.

Convergence between the cold-climate groups is predicted for the majority of morphological variables. Specifically, if the distinctive facial morphotype of the Greenland Eskimo is found to be cold-adapted, both the Argentinean and Canadian Groups (Fuèguian and Labradorean) will share features which result in the succinctly described "flat face" (Hanihara 2000; Hernández *et al.* 1997b; Lalueza *et al.* 1997). Populations from warm environments are predicted to have short upper facial height, basion-prosthion length and basion-bregmatic height, with shallow cheek heights. Maximum frontal, fronto-malar, minimum frontal, and maximum cranial breadths are predicted to be broad. Furthermore, bi-maxillary, bi-jugal, bi-zygomatic, and bi-condylar breadths will be narrow. Nasal height will be short, and nasal breadth, naso-dacryon subtense and inter-orbital breadth wide.

The Genetic Model
The Genetic Model (GM) is based on the evidence for waves of migration into the New World, assuming that all populations came from the same point of origin through the Bering Strait. Therefore, the assertion here is that the main adaptive pressure determining craniofacial characteristics is genetic. It represents the alternate hypothesis, predicting that the Greenland and Argentinean Groups (Angmaksaslik and Fuèguian) will show the largest degree of divergence of morphological traits, see Figure 2. If this model proves to be correct the groups will cluster in geographical sets. Therefore, there will be a three-way split, see Figure 2: the two Eskimo Groups (Angmaksaslik and Labradorean) in the far north forming cluster 1; the Californians and New Mexicans (Chumash and Ketchipauan) at an approximate geographical midpoint between the other sets forming cluster 2; and the Argentinean Group (Fuèguian) separated in the far south creating cluster 3. For this model the Eskimo Groups and the Argentinean Groups will show a large divergence of features. The heads of the Argentinean sample will be small and broad, produced by short upper facial height, basion-prosthion length and basion-bregmatic. Furthermore, maximum frontal, fronto-malar, minimum frontal, and maximum cranial breadths will be wide, and cheek height shallow. Mid-facial prognathism will not be present; therefore bi-maxillary breadth, bi-jugal breadth, bi-zygomatic breadth, and bi-condylar breadth will all be narrow. Finally, the nose will be broad and short as a consequence of short nasal height and broad nasal breadth, naso-dacryon subtense and inter-orbital breadth. If the traits are found to follow this model, the Californian and New Mexican populations' trait expression will demonstrate a resemblance to the Eskimo, Inuit and Argentinean Groups, without being as extreme as any of the cold-climate groups.

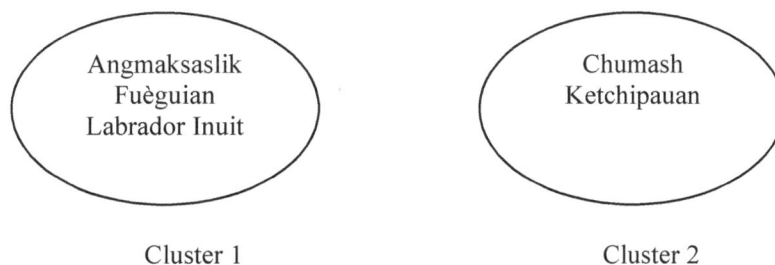

Figure 1. *Climatic Adaptation Model*

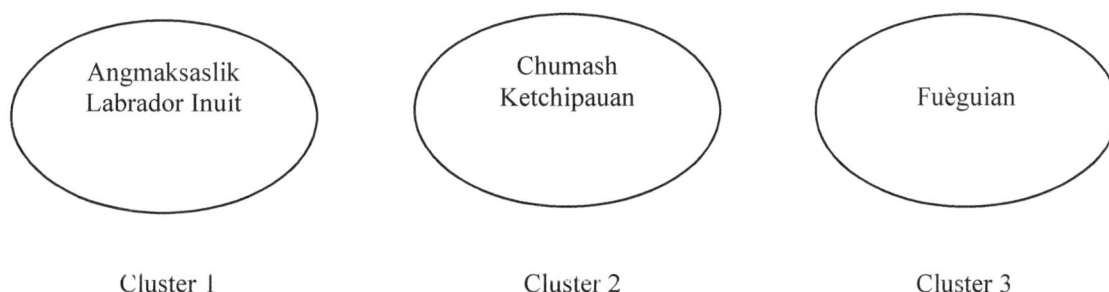

Figure 2. *Genetic Model*

TABLE 1. DEFINITIONS OF CRANIOFACIAL MEASUREMENTS

FACE = 12 MEASUREMENTS	ABBREV.	DESCRIPTION
BASION-PROSTHION LENGTH	BPL	Direct distance from (pr) to basion (ba) (Buikstra & Ubelaker 1997, 74).
NASION-PROSTHION HEIGHT	NPH	Facial height from nasion (n) to prosthion (pr).
NASAL HEIGHT, LEFT	NLH	The average height from nasion (na) to the lowest point on the border of the nasal aperture on the left side.
NASAL BREADTH	NLB	The maximum breadth of the nasal aperture (al-al) (Buikstra & Ubelaker 1997, 75).
NASO-DACRYON SUBTENSE	NDS	The subtense from the deepest point in the profile of the nasal bones to the interorbital breadth.
ORBITAL HEIGHT, LEFT	OBH	Direct distance between the superior and inferior orbital margins (Buikstra & Ubelaker 1997, 76).
ORBITAL BREADTH (LEFT)	OBB	Laterally sloping distance from dacryon (d) to ectoconchion (ec) (Buikstra & Ubelaker 1997, 76).
INTER-ORBITAL BREADTH	DKB	The direct distance between right and left dacryon (d-d) (Buikstra & Ubelaker 1997, 76).
BI-MAXILLARY BREADTH	ZMB	The breadth across the maxillae, from one zygo-maxillare anterior to the other (zyma-zyma).
BI-JUGAL BREADTH	JUB	The external breadth across the malars at the jugalia (j), i.e., at the deepest points in the curvature between the frontal and temporal process of the malars.
BI-ZYGOMATIC BREADTH	ZYB	The distance between the most lateral points on the zygomatic arches (zy-zy) (Buikstra & Ubelaker 1997, 74).
MINIMUM MALAR HEIGHT	WMH	The minimum distance, in any direction, from the lower border of the orbit to the lower margin of the maxillae, mesial to the masseter attachment on the left side.
MANDIBLE = 2 MEASUREMENTS		
BI-CONDYLAR BREADTH	bb	Direct distance between the most lateral points on the two condyles (cdl-cdl) (Buikstra & Ubelaker 1997, 78).
MANDIBULAR ANGLE	ma	Angle formed by the inferior border of the corpus and the posterior border of the ramus (Buikstra & Ubelaker 1997, 78).
VAULT = 5 MEASUREMENTS		
MAXIMUM CRANIAL LENGTH	GOL	Distance between glabella (g) and opisthocranion (op), in the midsagittal plane, measured in a straight line (Buikstra & Ubelaker 1997, 74).
MAXIMUM CRANIAL BREADTH	XCB	Maximum width of skull perpendicular to midsagittal plane (eu-eu) wherever it is located, with the exception of the inferior temporal lines and the area immediately surrounding them (Buikstra & Ubelaker 1997, 74).
BASION-BREGMATIC HEIGHT	BBH	Distance from bregma (b) to basion (ba).
MAXIMUM FRONTAL BREADTH	XFB	The maximum breadth at the coronal suture, perpendicular to the median plane.
MINIMUM FRONTAL BREADTH	WFB	Direct distance between the two fronto-temporale (ft-ft) (Buikstra & Ubelaker 1997, 75).
FRONTO-MALAR BREADTH	FMB	The most anterior point the fronto-malare suture, measured directly between two external points (fmt-fmt).

TABLE 2. DEFINITIONS OF CRANIOFACIAL INDICES

FACIAL SKELETON: 3 INDICES	VALUE
1 UPPER FACIAL INDEX	$\dfrac{\text{Upper Facial Height (NPL)} \times 100}{\text{Bi-Zygomatic Breadth (ZYB)}}$
2 NASAL INDEX	$\dfrac{\text{Nasal Breadth (NLB)} \times 100}{\text{Nasal Height (NLH)}}$
3 ORBITAL INDEX	$\dfrac{\text{Orbital Height (OBH)} \times 100}{\text{Orbital Breadth (OBB)}}$
VAULT : 4 INDICES	
4 CRANIAL INDEX	$\dfrac{\text{Maximum Cranial Length (GOL)} \times 100}{\text{Maximum Cranial Breadth (XCB)}}$
5 CRANIAL LENGTH-HEIGHT INDEX	$\dfrac{\text{Basion-Bregmatic Height (BBH)} \times 100}{\text{Maximum Cranial Length (GOL)}}$
6 CRANIAL BREADTH-HEIGHT INDEX	$\dfrac{\text{Basion-Bregmatic Height (BBH)} \times 100}{\text{Maximum Cranial Breadth (XCB)}}$
7 FRONTO-PARIETAL INDEX	$\dfrac{\text{Minimum Frontal Breadth (WFB)} \times 100}{\text{Maximum Cranial Breadth (GOL)}}$

Materials

This study was carried out using a selection of craniofacial measurements derived from five modern geographically and craniometrically diverse populations. The majority of data was collected by the author, although data was also included that was collected by Dr. M. Lahr (Lahr 1996). The definitions of the 20 craniofacialmetric variables and 7 indices selected are given in Tables 1 & 2. All specimens consisted, where possible, of virtually complete skulls. The main selection criterion was the locality of the original population, although the genetic relationships and evolutionary histories of the population groups were also significant factors. Further selection criteria related to the temperature of the area from which the populations originated: three of the five populations selected were chosen for their cold climate, and two for their warm climate. The selection of population groups from both southern and northern regions of the Americas allows for trends in trait expression between these groups to be assessed when compared with groups from regions in-between the geographical extremes. The inclusion of these intermediate groups will offer invaluable information necessary for this study with regards to

extraneous factors affecting the traits, such as long-term genetic influences.

Population Samples

The most northerly group selected for study are the Eskimos of Greenland. This sample comprises 33 individual crania; 14 male, 19 female. The majority of the skulls originated from south-eastern Greenland, specifically Angmaksaslik which is an historic site located on the east coast. Culturally the Greenland Eskimo are very well adapted to cold. Characteristic features usually include a dolichocephalic vault, pronounced sagittal keeling, gracile supraorbital morphology, large orbits, small nasal apertures and nasal bones, pronounced facial flatness, large faces with tall and broad malars, high temporal lines, thick tympanic bones, palatal tori, and M3 agenesis (Lahr 1995). They are also relatively gracile. The climate can be described as cold, with varying amounts of rainfall (Binford 2001).

The second most northerly group selected are the Eskimos of Northern Labrador. Although this sample is numbered as Group 4, as it is a supplementary set, it is arranged with the other cold-climate groups for clarity of comparison. The Labrador sample comprises 17 individual crania; 9 male and 8 female. All of the specimens in this sample originate from north-eastern Labrador, off the coast of Newfoundland. Historical records at the Natural History Museum state that the remains are likely to be more than 300 years old, and represent the remains of the Eskimo population that inhabited the area. Culturally the Labradorean Inuit are as well adapted to cold as other Inuit groups. The Labradorean climate can be described as cold, with average rainfall (Binford 2001). Figures 3 and 4 show a male Labradorean, sample number AM.1.0.1.

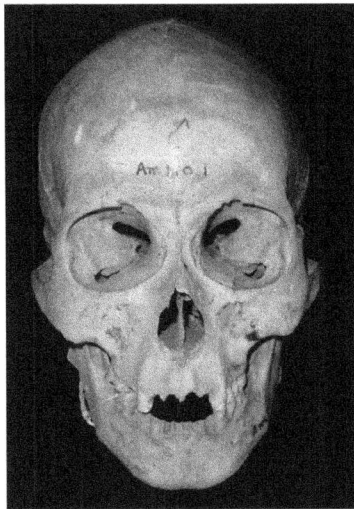

Figure 3. Male Labradorean Eskimo, anterior view. Sample number AM.1.0.1

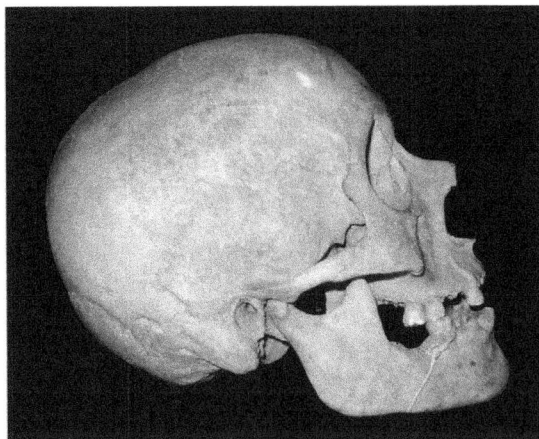

Figure 4. Male Labradorean Eskimo, lateral view. Sample number AM.1.0.1

Group 2 originates from Tierra del Fuègo, consisting of members of the human groups found at the southernmost latitude of Argentina. The climate can be described as cold, with varying amounts of rainfall (Binford 2001). The Argentinean sample comprises 33 individual crania; 16 male, 5 female, and 12 unsexed. The first impression obtained when looking at the sample is one of robusticity and large size. The skulls show very pronounced

supraorbital ridges, which are fully modern in shape and possibly the most pronounced of any modern crania. Keeling of the vault is also a very common feature, although the most characteristic feature is the presence of occipital tori. In terms of cranial dimensions, besides the absolute large size, the material from Tierra del Fuègo can be characterised by a very tall and broad face (Lahr 1995).

The most westerly group included in this study are the Santa Cruz Islanders, of the east coast of California. The sample comprises 33 individual crania, 20 of which were sexed as male, 13 female. Although the climate was cool and dry, the Chumash are labelled warm given the relative climatic warmth versus the 'cold' groups. Figures 5 and 6 show an example of a male Chumash Indian, sample number Anth42/3.

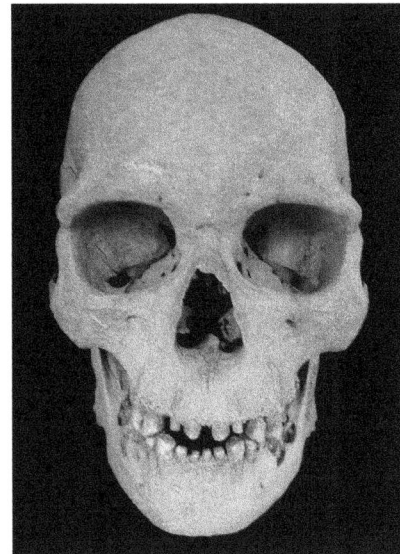

Figure 5. Male Chumash Indian, anterior view

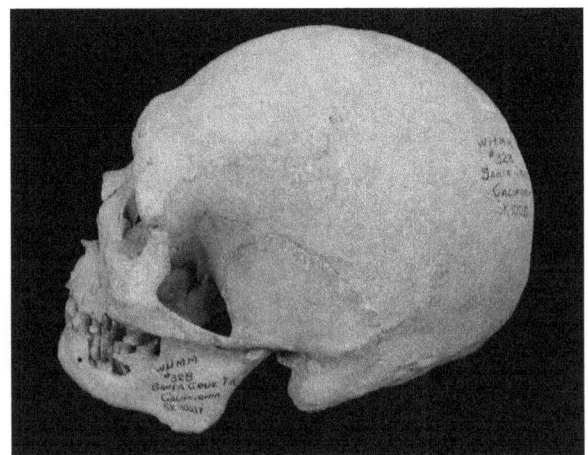

Figure 6. Male Chumash Indian, lateral view

The most central group selected for analysis are the native New Mexican Pueblo Indians, known as the

Ketchipauans. The sample comprises of 20 individual crania; 7 male and 12 female, 1 unsexed. Although the Pueblo Indian geographical range was large, the entire sample originates from an open pueblo situated on a summit 24 km from Zuni Pueblo in Western Central New Mexico. One of the most important characteristics of precipitation in the Ketchipauan area is its variability from year to year, regardless of elevation. The Pueblo Indians are known to have carried out artificial cranial deformation, the results of which may influence some aspects of the morphometric analyses. Despite this, they will be included in this study as cranial deformity primarily influences the morphology of the vault (Anton 1989). The Ketchipauan samples therefore may still provide information regarding the relationship between facial morphology and climate.

Methods
Testing the Hypotheses
The Climatic Adaptation and Genetic Models represent the predicted pattern of variable results for each of the populations. They are based on the features associated with theories of cold-climate adaptation in the craniofacial skeleton (Howells 1973, 1989; Lahr 1995, 1996; Steegman 1970a, 1975; Steegman et al. 2002) and the theory of regional continuity in the selected populations. In this way, a comparative foundation was produced that could be used to develop as well as test models at a later stage. Both models are presented in Table 3, simplified to represent the suite of traits expected as the prediction in both cases is convergence between various population groups. The variables of mandibular angle (Ma), maximum cranial length (GOL), and orbital height (OBH) and breadth (OBB) were not included as they are not predicted to be influenced by climate. They are incorporated into the study, however, as they are necessary to form an overall understanding of the variability in craniofacial morphology of the populations.

TABLE 3. PREDICTIONS OF CLIMATIC ADAPTATION & GENETIC MODELS

VARIABLE	CLIMATIC ADAPTATION MODEL		GENETIC MODEL		
	COLD-CLIMATE	WARM-CLIMATE	FUÈGUIAN	CHUMASH & KETCHIPUAUA	ANGMAKSASLIK & LABRADOREAN INUIT
	C	W	C	W	C
BPL	L	S	S	I	L
NPH	L	S	S	I	L
NLH	L	S	S	I	L
NLB	N	B	B	I	N
NDS	S	L	L	I	S
OBH	L	x	x	x	L
OBB	L	x	x	x	L
DKB	N	B	B	I	N
ZMB	B	N	N	I	B
JUB	B	N	N	I	B
ZYB	B	N	N	I	B
WMH	D	SH	SH	I	D
Bb	B	N	N	I	B
Ma	x	x	x	x	x
GOL	L	S	S	I	L
BBH	L	S	S	S	L
XFB	N	B	B	I	N
FMB	N	B	B	I	N
WFB	N	B	B	I	N
XCB	N	B	B	I	N

L	=	Long	S	=	Short
B	=	Broad	N	=	Narrow
D	=	Deep	SH	=	Shallow
I	=	Intermediate	x	=	No predictive quality
C	=	Cold-climate	W	=	Warm-climate

Statistical Analysis

The aim of the analyses was to determine if the Angmaksaslik and Fuèguian Groups are more similar to each other in their craniofacial morphology, in terms of both size and shape, than the population groups located geographically between them. To this end, the regional distribution of each of the metrical variables and indices were initially tested using one-way analysis of variance (ANOVA). Following this, Canonical Variates Analysis (CVA) was utilised to investigate trends in the data. Although the mandibular measurements were initially incorporated in this study, following a number of CVAs, their inclusion was found to limit the data set too severely as this removed the entire Labradorean Intuit Group because no mandibles were present with any of the samples. Subsequently, mandibular variables were removed from all further analyses.

ANOVA

Size and shape variability among the population samples was analysed using ANOVA; Table 4 provides details of the model supported by each variable and indices (see Figures 7-10). From Table 4 it can be seen that 10

variables strongly supported the CAM, a further 3 follow the predictions of the CAM in part, 2 follow the GM, and the results of 4 do not support either model. For the facial complex, 10 of the possible 12 variables follow the CAM, and the remaining 2 follow neither model. The cranium complex is a mixture of variables, 3 of which support the CAM, 2 the GM (the only 2 variables or indices to support this model), leaving 1 that does not follow either predictive pattern. When looking at shape variation, only orbital index supported the predictions of a model, that of CAM.

Of all the analyses undertaken, four examples of relationships were found between the populations in different variables, shown in Figures 7-10. Numbers 1 & 2 relate to the Climatic Adaptation Model, although the results are not fully consistent as the Labrador Inuit cluster with the other warm groups, and example 3 relates to the Genetic Model. As Table 4 demonstrates, case 1 is the most common, as 4 variables exhibit this relationship between samples. No other pattern was replicated exactly, other than the exclusion of the Labradorean Inuit group from all others, whilst the remaining four populations form a single cluster, as shown in case 4.

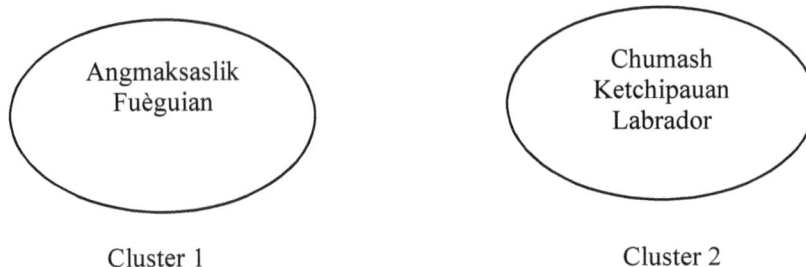

Angmaksaslik
Fuèguian
Cluster 1

Chumash
Ketchipauan
Labrador
Cluster 2

Figure 7. Climatic Adaptation Model, Number 1

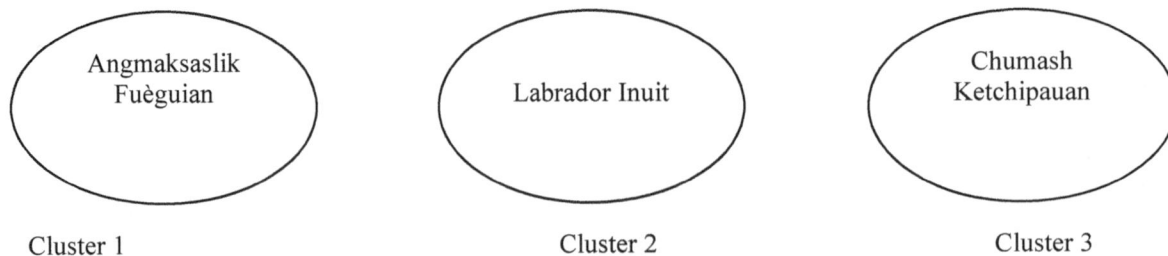

Angmaksaslik
Fuèguian
Cluster 1

Labrador Inuit
Cluster 2

Chumash
Ketchipauan
Cluster 3

Figure 8. Climatic Adaptation Model, Number 2

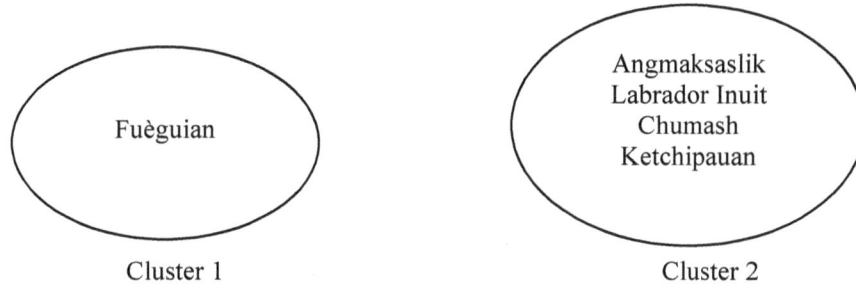

Figure 9. Genetic Model, Number 3

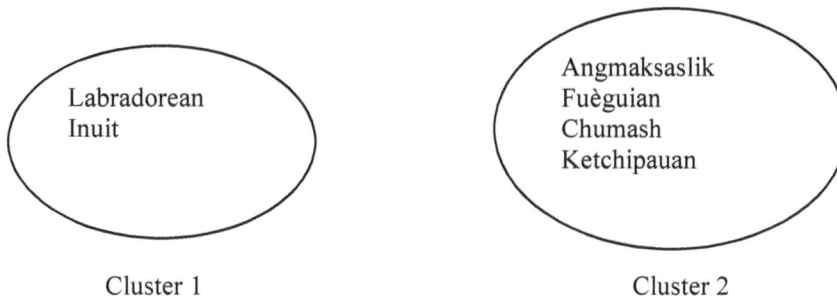

Figure 10. Labradorean Inuit Removal, Number 4

TABLE 4. SUMMARY OF VARIABLES AND INDICES: CAM1 (CLIMATE MODEL 1); CAM 2 (CLIMATE MODEL 2); GM (GENETIC MODEL, 3); LABRADOR INUIT REMOVAL, MODEL 4

CRANIOFACIAL MEASUREMENTS	MODEL SUPPORTED
FACIAL VARIABLES = 12	
BASION-PROSTHION LENGTH	CAM 1
UPPER FACIAL HEIGHT	CAM 1
NASAL HEIGHT, LEFT	CAM 1
NASAL BREADTH	-
NASO-DACRYON SUBTENSE	LAB 4
ORBITAL HEIGHT, LEFT	CAM 2
ORBITAL BREADTH (LEFT)	CAM 1
INTER-ORBITAL BREADTH	CAM
BI-MAXILLARY BREADTH	CAM 4
BI-JUGAL BREADTH	CAM
BI-ZYGOMATIC BREADTH	CAM
MINIMUM MALAR HEIGHT	CAM
VAULT VARIABLES = 6	
MAXIMUM CRANIAL LENGTH	-
MAXIMUM CRANIAL BREADTH	CAM
BASION-BREGMATIC HEIGHT	CAM
MAXIMUM FRONTAL BREADTH	GM
MINIMUM FRONTAL BREADTH	CAM
FRONTO-MALAR BREADTH	GM 3
CRANIOFACIAL INDICES	
FACIAL INDICES = 3	
UPPER FACIAL	-
NASAL	-
ORBITAL	CAM
VAULT INDICES = 4	
CRANIAL	-
CRANIAL LENGTH-HEIGHT	-
CRANIAL BREADTH-HEIGHT	-
FRONTO-PARIETAL	-

Canonical Variates Analysis
Following the use of ANOVAs, the variables from the five population groups were analysed using Canonical Variates Analysis (CVA), as this type of statistical analysis has the capacity to achieve maximum divergence between variables, facilitating the determination of trends of convergence.

In order to examine the overall effects of craniofacial size, an analysis was undertaken on all variables. In this model, Analysis 1, of the 136 cases available 28 (20.6%) had at least one missing variable. Consequently, 108 (79.4%) were used in the analysis. Of the original grouped cases 72.7% were correctly classified. On function 1 the Labrador set is removed from all others as a result of maximum cranial length and naso-dacryon subtense. These variables are not correlated (r = 0.059), therefore they are independently contributing to the removal of the Labrador sample. On function 2, maximum cranial length is responsible for the removal of the Ketchipauan sample from the single cluster group, which is formed by the Angmaksaslik, Fuèguian, Chumash sets.

Analysis 2 consists of all facial variables, thus analyzing facial size (see Figure 11). From this figure, it can be seen that on function 1, Labrador is removed from all other groups. This is a result of naso-dacryon subtense and bi-jugal breadth, as the correlation result for these two variables is low (r = 0.1). On function 2, nasal height has the most discriminatory power. Therefore, the two clusters seen in Figure 11 are the result of all three variables; the first cluster contains the two cold-climate groups, the Angmaksaslik Eskimo and Fuèguian Indians, the second cluster consists of the warm-climate Chumash and Ketchipauan samples.

Analysis 3 was undertaken on the indices derived from the variables that related to facial shape. In total, 3 of the 7 indices were included in this investigation. This analysis omitted 26 (19.1%) cases, as a result of one or more missing variables. Subsequently, 110 (80.9%) of the possible 136 cases were included. On function 1, nasal index causes general scattering of the populations, with the Angmaksaslik and Ketchipauan samples forming the two extremes, and the remaining populations being distributed between them. Orbital index on function 2 is responsible for limited spreading of the samples, with the Ketchipauan set being the only group removed on the Y-axis. A single cluster is formed containing the Fuèguian, Chumash, and Labradorean samples, the group centroids of which are very close together.

Discussion
This study was designed to explore the variability of the craniofacial features hypothesised to correspond to adaptation to cold climates, and to determine whether they co-vary with each other and/or with climatic gradients. The populations studied consisted of five groups of indigenous Native Americans who lived in areas with different climatic conditions.

The univariate results of size variation illustrate that naso-dacryon subtense, bi-condylar breadth, and minimum frontal breadth show significant differences (>3mm) between the groups. In total, 13 of the 20 variables are consistent with the CAM, although the Labrador sample did not cluster with the other two cold-climate groups. The following variables further divide the samples and show morphological similarity between the Angmaksaslik and Fuèguian groups (cold-climate), as well as the Chumash and Ketchipauan samples (warm-climate): basion-prosthion length, upper facial height,

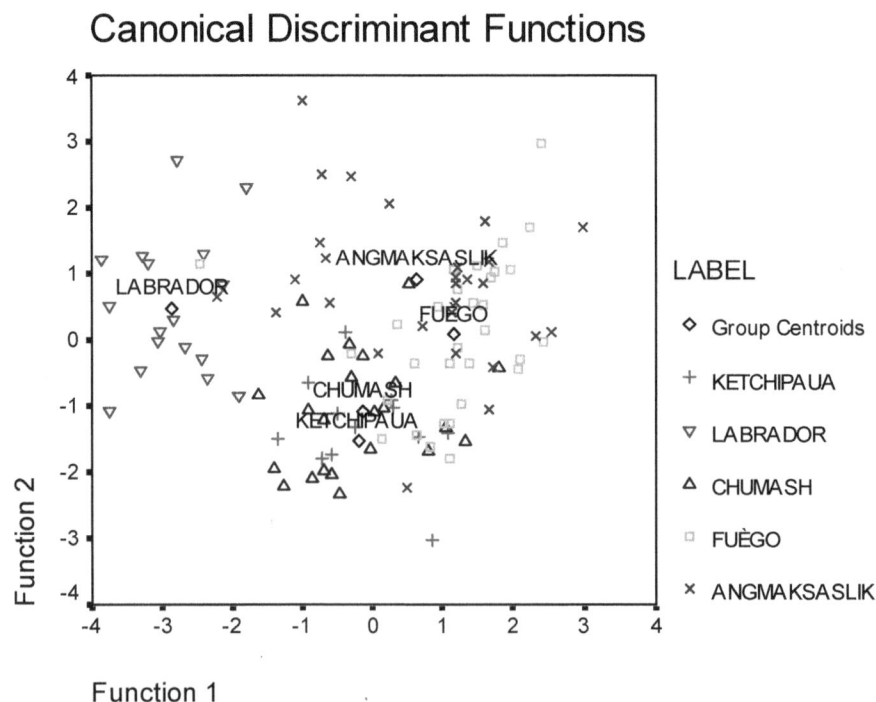

Figure 11. Facial Variables

10

nasal height, orbital height and breadth, inter-orbital breadth, bi-maxillary breadth, bi-jugal breadth, bi-zygomatic breadth, maximum malar height, maximum cranial breadth, basion-bregmatic height, and minimum frontal breadth. The two cold-climate groups differ significantly in maximum frontal breadth and fronto malar breadth, both of which are shorter than expected in the Fuèguian sample.

When the results are analysed in terms of long-term adaptation to environment they suggest that a number of features of both cold- and warm-climate samples follow the CAM, as the craniofacial morphology of the Chumash and Ketchipauan Indians differs from that of the Angmaksaslik and Fuèguians in the ways predicted by this model. For example, the Chumash and Fuèguian Indians' upper facial height, basion-prosthion length, and basion-bregmatic height are short, and cheek height is shallow, representing a short head and face. Furthermore, the measurements of bi-jugal and bi-zygomatic breadths demonstrate that mid-facial prognathism is not present. However, the Chumash and Ketchipauan faces were not narrow, a result inconsistent with the CAM, as maximum frontal breadth, fronto-malar breadth and minimum frontal breadth were not short as predicted. Similarly, the Fuèguian sample did not follow the CAM in relation to facial breadth, as their faces were not found to be wide. This suggests that the craniofacial adaptation of both the warm- and cold-climate groups is affected by climatic pressure in approximately the same traits. However, as the majority of variables follow the prediction laid out in the CAM, except facial width, the results for the univariate size analyses support the hypothesis that overall craniofacial size is associated with climatic influence in the population groups included in this study.

The CVA relating to size variation among the samples show that the two variables of the most discriminant power were maximum cranial breadth and naso-dacryon subtense; interesting as neither variable has previously been suggested as important when investigating population variability, or linked with climatic adaptation. Of the models produced, Analysis 1, which included all variables, supported the GM as the Fuèguian and Angmaksaslik samples cluster around the Chumash Group. Analysis 2, facial variables, demonstrated a cluster pattern supportive of the CAM. This indicates that only facial size is linked to climatic pressure, although in these models the warmer climate samples did not cluster to support the hypothesis of climatic influence on head and face size. Additionally, the results of the facial analysis, Analysis 2, are important as they are unlikely to have been influenced by the Ketchipauan cranial deformation. The results for all analyses were supported by the single-sex tests, signifying that the observed morphological patterns do not affect males and females differentially.

It can be seen from these results that both univariate and canonical analyses relating to size largely correspond with the Climatic Adaptation Model, a conclusion not supported by previous studies which used similar types of data. Therefore, cold-climate adaptation appears to be largely responsible for the overall size of the faces of both the cold- and warm-climate populations analysed.

The results for the univariate shape analysis show that orbital index, nasal index, cranial index, and cranial breadth-height index, and fronto-parietal index are of high discriminatory value when differentiating the samples. As the results for the Angmaksaslik and Fuèguian groups show comparability for orbital index, as do the Chumash and Ketchipauan samples, this suggests that orbit shape is related to cold-climate adaptation. The patterns of cluster grouping seen in the univariate results do not correspond with the hypothesis laid out in the GM.

With regard to the canonical shape variation analyses, no clustering is found between the cold- or warm-climate groups. However, when the description of the ranges into which the values fall are analysed, only nasal index is of discriminant value when differentiating cold-climate from warm-climate groups. This supports the conclusion of previous authors that nasal proportions are associated with cold-climate adaptation (Steegman 1970a), although this index is not found to be diagnostic in this study.

From the results it can be seen that in both univariate and canonical analyses, orbital index is the only index found to correspond to cold-climate adaptation. This suggests that, although climatic adaptation may not have been the primary adaptive pressure on the overall shape of craniofacial morphology of the samples, it appears to have been a significant influence on the shape of the eye socket, in terms of both height and width.

Predicted vs. Attained Results
A number of features are shared between the Eskimo and Fuèguian samples, and they are found to be similar in 14 craniofacial dimensions. The Labradorean Inuit sample, however, differs from the other cold-climate groups in all craniofacial measurements, except nasal breadth.

As the Labrador set is removed from the other cold-climate groups in every analysis this suggests that one or more extraneous factors are affecting the size and shape of the Labradorean heads. That the Labradorean Inuit sample varies greatly from the predicted results is of particular interest as many studies combine such groups under the generic term "Eskimo". The results of this study suggest that this method could lead to the loss of important diagnostic information, potentially resulting in erroneous statistical results. Without the comparison of the Chumash and Ketchipauan samples, the Labradorean Inuit's results could imply that the correlation in features between the Angmaksaslik and the Fuèguian sets are due to factors other than cold-climate adaptation. However, that the Angmaksaslik and Fuèguian groups cluster supports the CAM, a pattern reflected in the morphology of the warm-climate samples providing further evidence of climatic influence on overall craniofacial morphology.

Conclusion

This study has illustrated the presence of a relationship between a number of aspects of craniofacial size and shape and climatic conditions among five groups of Native Americans. Climate appears to have influenced the craniofacial skeleton in the majority of variables included, demonstrating that cold-climate selection is of great importance on the overall size of cold-climate heads and faces. Of the variables that suggest this, convergence of maximum cranial length and orbital breadth between the Angmaksaslik and Fuèguian Groups, as well as the Chumash and Ketchipauan samples, indicates that these two variables should be included in studies assessing adaptation to environment.

The results for shape variation are more limited with regard to indicating climatic pressure on the craniofacial skeletons of the samples studied, and suggest that only orbital index is associated with cold-climate adaptation. Nasal index was not found to be of high discriminatory value, a conclusion which is inconsistent with the results of earlier studies (Franciscus 2003; Hernández et al. 1997a; Steegman 1970a; Thomson 1913). However, this pattern of trait expression is a probable result of cold-climate adaptation as, in contrast, the warmer-climate groups (the Chumash and Ketchipauan Indians) showed shorter maximum cranial length and wider nasal breadth. Although the observed morphological variation is the likely result of climatic adaptation pressure on the craniofacial skeletons of the populations analysed, as it is both consistent and statistically significant, it is not possible on the basis of the available evidence to rule out biomechanical adaptation as a cause of the observed morphological variability. However, regardless of the source of this variability, observed craniofacial differences follow patterns of morphology that would be predicted from climatic data. This suggests that there is potential for climate-led morphological analyses to be used to infer processes of adaptation to environment. The remaining results for the indices representing shape variability do not correspond with the genetic model, and are therefore not predominantly the result of genetic history or cold-climate selection.

Although the classification results in this study are good, in order to get a true understanding of how well the proposed classification system performs new cases require testing with this method, as only the classification of new cases allows the predictive validity of the classification functions. The patterns of trait expression demonstrated here could be compared to many modern world population groups outside the Americas, as well as archaic remains. Therefore, although this study offers evidence in support of climatic factors influencing the morphology of the craniofacial skeleton, further work on statistical models that directly incorporate temperature is required to support the conclusions presented here. One method would be to subject each variable to univariate analysis, with population and effective temperature as factors; this will offer information on whether population or climate explains more of the variability in each measurement or indices. It would also prove interesting if

similar models could be devised for the post-cranium and tested on the populations in this study where possible, to permit the comparison of the craniofacial and postcranial remains.

Finally, the hypothetical models used to predict morphology were developed for this investigation, and have therefore not been applied to other population groups. Consequently, it may be of importance to re-analyse the Neanderthal cold-environment adaptation hypothesis with the results of this study in mind, due to the body of contention that surrounds cold-adaptation in general as well as adaptive pressures on the morphology of the Neanderthals.

Acknowledgements
I would like to thank Dr. Jay Stock, Dr. Lucio Castilho, Dr. Marta Lahr, & Maggie Bellatti from the University of Cambridge, and Louise Humphrey and Robert Kruszynski from the British Natural History Museum. I would also like to thank Nick Ray for his support.

References

Adams JM, Foote GR, Otte M. 2001. Could Pre-Last Glacial Maximum Humans Have Existed in North America Undetected? An Interregional Approach to the Question. *Current Anthropology* 42(4):563-566.

Anton SC. 1989. Intentional Cranial Vault Deformation and Induced Changes of the Cranial Base and Face. *American Journal of Physical Anthropology* 79:253-267.

Biggers JD, Ashoub MR, McLaren A, Michie D. 1958. The Growth and Development of Mice in Three Climatic Environments. *Journal of Experimental Biology* 35:144-155.

Binford LR. 2001. *Constructing Frames of Reference: an analytical method for archaeological theory building using ethnographic and environmental data sets.* University of California Press, London.

Blumenfeld J. 2002. Neanderthal Facial Morphology and Cold Adaptation: a comparative approach. *American Journal of Physical Anthropology* 34 Annual Meeting (Supp.):45.

Buikstra JE & Ubelaker DH (editors). 1997. *Standards: for data collection from human skeletal remains.* Arkansas Archaeological Survey, Arkansas.

Churchill SE. 1998. Cold Adaptation, Heterochrony, and Neanderthals. *Evolutionary Anthropology* 7(2):46-61.

Colvin W. 2002. *History of Santa Cruz Island* www.ic.ucsc.edu/~envs160/scisland.htm (07.05.03)

Coon CS. 1971. *The Story of Man.* 3rd ed. Alfred A. Knoft, New York.

Foley RA. 1994. Speciation, Extinction and Climatic Change in Hominid Evolution. *Journal of Human Evolution* 26:275-289.

Fox CL, Hernández M, Moro CG. 1996. Craniometric Analysis in Groups from Tierra del Fuego/Patagonia and the Peopling of the South Extreme of the Americas. *Human Evolution* 11(3-4):217-224.

Franciscus RG. 2003. Internal Nasal Floor Configuration in *Homo* with Special Reference to the Evolution of Neanderthals Facial Form. *Journal of Human Evolution* 44:701-729.

Friess M. 2002. Revisiting Human Cold Adaptation: craniofacial shape assessed by 3D laser scanning. *American Journal of Physical Anthropology* 34(Annual Meeting (Supp.)):72.

González-José R, Dahinten SL, Luis MA, Hernández M, Pucciarelli HM. 2001. Craniometric Variation and the Settlement of the Americas: testing hypotheses by means of R-matrix and matrix correlation analyses. *American Journal of Physical Anthropology* 116:154-165.

Greenberg JH, Zegura SL. 1986. The Settlement of the Americas: a comparison of the linguistic, dental, and genetic evidence. *Current Anthropology* 27(5):477-497.

Guglielmino-Matessi C, Gluckman P, Cavalli-Sforza L. 1979. Climate and the Evolution of Skull Metrics in Man. *American Journal of Physical Anthropology* 50:549-564.

Hanihara T. 1979. Dental Traits in Ainu, Australian Aborigines, and New World Populations. In *The First Americans: origins, affinites, and adaptations*, edited by W. S. Laughlin and A. B. Harper, pp. 125-134. Gustav Fischer, New York.

Hanihara T. 1993a. Population Prehistory of East Asia and the Pacific as Viewed from Craniofacial Morphology: the basic populations in East Asia, VII. *American Journal of Physical Anthropology* 91:173-187.

Hanihara T. 1993b. Craniofacial Features of Southeast Asians and Jomonese: a reconsideration of their micro-evolution since the late Pleistocene. *Anthropological Science* 101:25-46.

Hanihara T. 1996. Comparison of Craniofacial Features of Major Human Groups. *American Journal of Physical Anthropology* 99:389-412.

Hanihara T. 2000. Frontal and Facial Flatness of Major Human Populations. *American Journal of Physical Anthropology* 111:105-134.

Hernández M, Fox CL, Garcia-Moro C. 1997a. Fueguian Cranial Morphology: the adaptation to a cold, harsh environment. *American Journal of Physical Anthropology* 103(103-117).

Howells WW. 1960a. *Mankind in the Making*. Secker and Warburg, London.

Howells WW. 1960b.Criteria for Selection of Osteometric Dimensions. *American Journal of Physical Anthropology* 30:451-458.

Howells WW. 1973. *Cranial Variation in Man: a study by multivariate analysis of patterns of difference among recent human populations*. Papers of the Peabody Museum of Archaeology and Ethnography 67. Harvard University, Massachusetts.

Howells WW. 1989. *Skull Shapes and the Map: craniometric analyses in the dispersal of modern Homo*. Papers of the Peabody Museum of Archaeology and Ethnography 79. Harvard University, Massachusetts.

Hylander WL. 1977. The Adaptative Significance of Eskimo Craniofacial Morphology. In *Orofacial Growth and Development*, edited by A. A. Dahlberg and T. M. Graber, pp. 129-169. Mouton, The Hague.

Lahr MM. 1995. Patterns of Modern Human Diversification: implications for Amerindian origins. *Yearbook of Physical Anthropology* 38:163-198.

Lahr MM. 1996. *The Evolution of Modern Human Diversity: a study of cranial variation*. Cambridge University Press: Cambridge.

Lahr MM & Wright RVS. 1996. The Question of Robusticity and the Relationship Between Cranial Size and Shape in *Homo sapiens*. *Journal of Human Evolution* 31:157-191.

Lalueza C, Hernandez M, Garcia-Moro C. 1997. La Morfologia Facial de las Poblaciones Fueguinas: Êreflejo de una adaptacion al frio? *Ans. Inst. Pat. Ser. Cs. Hs. Punta Arenas (Chile)* 25:45-58.

Lieberman DE, Pearson OM, Mowbray KM. 2000. Basiocranial Influence on Overall Cranial Shape. *Journal of Human Evolution* 38:291-315.

Neves WA & Pucciarelli HM. 1991. Morphological Affinities of the First Americans: an explanatory analysis based on early South American human remains. *Journal of Human Evolution* 21:261-273.

Ogle C. 1934. Climatic Influence on the Growth of the Male Albino Mouse. *American Journal of Physiology* 107:635-640.

Powell JF & Neves WA. 1999. Craniofacial Morphology of the First Americans: pattern and process in the peopling of the New World. *Yearbook of Physical Anthropology* 42:153-188.

Steegman AT. 1965. A Study of Relationships Between Facial Cold Responses and Some Variables of Facial Morphology. *American Journal of Physical Anthropology* 23:355-362.

Steegman AT. 1970a. Cold Adaptation and the Human Face. *American Journal of Physical Anthropology* 32:243-250.

Steegman AT. 1972. Cold Response, Body Form, and Craniofacial Shape in Two racial Groups of Hawaii. *American Journal of Physical Anthropology* 37:193-222.

Steegman AT. 1975. Human Adaptation to Cold. In *Physiological Anthropology*, edited by A. Damon, pp. 130-166. Oxford University Press, New York.

Steegman AT, Cerny FJ, Holliday TW. 2002. Neanderthal Cold Adaptation: physiological and energetic factors. *American Journal of Human Biology* 14:566-583.

Steegman AT & Platner WS. 1968. Experimental Cold Modification of Cranio-facial Morphology. *American Journal of Physical Anthropology* 28:17-30.

Steele J, Adams J, Sluckin T. 1998. Modelling Paleoindian Dispersals. *World Archaeology* 30(2):286-305.

Summer FB. 1909. Some Effects of External Conditions upon the White Mouse. *Journal of Experimental Zoology* 7:97-155.

Thomson A. 1913. The Correlations of Isotherms with Variations in the Nasal Index. *International Congress of Medicine, London* 2:89-90.

Vrba E. 1980. Evolution, Species and Fossils: how does life evolve? *African Journal of Science* 76:61-84.

Walker PL. 1996. AAPA Statement on Biological Aspects of Race. *American Journal of Physical Anthropology* 101:569-570.

Wolpoff MH. 1968. Climatic Influence on the Skeletal Nasal Aperture. *American Journal of Physical Anthropology* 29:405-423.

CANOPY HEIGHT UTILISATION AND TRAUMA IN THREE SPECIES OF CERCOPITHECOID MONKEYS

C. Chapman, S.S. Legge, S.E. Johns

Abstract

Trauma was studied in the long bones (femur, tibia, fibula, humerus, radius and ulna) of primate skeletal remains housed at the Powell-Cotton Museum, Birchington, Kent, UK. The specimens were from three different arboreal quadrupeds who are known to travel at overlapping but differential levels in the tree canopy; Cercopithecus nictitans, Cercopithecus cephus, and Piliocolobus badius. Of the 80 skeletons examined, 15 had evidence of healed fractures. The femur was found to be the most frequent type of bone broken amongst all three species, followed by the humerus, then radius. Although there was a trend toward a higher trauma frequency based on increased canopy travelling height, the differences were not statistically significant.

Keywords: Trauma; fracture; canopy; primate; cercopithecoid

Introduction

Previous research has indicated that healed trauma in non-human primates is relatively common. Schultz (1956a) suggests that between 20 and 30% of individuals in most primate groups have healed fractures. Lovell (1991) also states that the available evidence suggests that one in five animals display at least one healed fracture. However, most of the studies have tended to concentrate on the great apes (Duckworth 1911; Jurmain 1989; Lovell 1990; Schultz 1956b) with relatively few studies on the more numerous smaller primates (Bramblett 1967; Buikstra 1975; Lovell 1991; Nakai 2003; Schultz 1956b).

Additionally, primate trauma studies are extremely diverse in their methods of analysis, sample size, and geographic and temporal origins of the specimens.

Therefore, there is a need to standardize the collection and reporting of primate pathological data to enable accurate analysis and to determine meaningful associations between the percentage of healed fractures and factors such as subsistence, locomotion, and social behaviours (Lovell 1991). Further, rather than the study of healed fractures in their own right or using fracture data to reconstruct life histories (Nakai 2003), the results are often placed within an evolutionary framework of analysis. For instance, both Lovell (1991) and Schultz (1937, 1956a, 1956b) were concerned with how healed fractures affected reproductive fitness.

Trauma among non-human primates may be the result of a variety of causes, including inter- or intra-species violence or falls from the tree canopy. One particular study has indicated that arboreal primates have a much greater chance of sustaining trauma than do terrestrial species (Preuschoft 1990). Following from Preuschoft's observations, and Lovell's (1991) inference that trauma may be related to arboreality, this preliminary study examines primate skeletal trauma in the context of travelling height. Specifically, we investigate whether arboreal quadrupeds who utilise the higher levels of the canopy for travelling sustain a greater frequency of skeletal trauma in the long bones than those who utilise the lower levels.

Materials and Methods

The three primates chosen for this study were *Cercopithecus nictitans, Cercopithecus cephus,* and *Piliocolobus badius.* The specimens were all collected in Cameroon, West Africa between the years of 1932-39 and are currently accessioned at the Powell-Cotton Museum in Birchington, Kent, UK. Between 25 and 28 specimens were examined for each species (Table 1).

Table 1. Number of specimens examined by species.

Species	Male	Female	Total
Cercopithecus cephus	21	6	27
Cercopithecus nictitans	21	4	25
Piliocolobus badius	7	21	28

Trauma data was only collected for the long bones of the specimens observed because it is often difficult to distinguish whether injuries to the digits and cranium are caused by violence rather than falling (Lovell 1991). Data on fractures was collected through macroscopic analysis. Each bone was compared with its counterpart for every individual specimen (i.e. left ulna with right ulna) to examine whether there were any morphological differences between the sides of the body. In this way fractures resulting in subtle morphological changes, which may not have been apparent had the bones been examined individually, were identified (see Figures 1 and 2). Fractures were diagnosed via callus formation and fracture complications were diagnosed from observation and from photographs taken of the skeletal material at Powell Cotton and compared to photographs and drawings of similar type fractures and fracture complications (Ortner 2003; Schultz 1937, 1956a, 1956b; White 2002), Additional observations regarding the fracture included: bone involved; location of the fracture on the bone and when possible, the type of force involved in creating the fracture. The study included only antemortem healed and healing fractures. Perimortem and postmortem fractures were noted but disregarded in the final analysis as it was unclear how these fractures were obtained. While this study focused on fractures, all pathological bone alterations associated with the injuries were also described, such as osteoarthritis, bone remodelling, traumatic myositis ossificans and abnormal fusion.

A literature review was undertaken to determine the travelling height within the canopy utilised by each species. This proved to be difficult as numerous factors may affect canopy usage, even within the same species. These could include the utilisation of different substrates, differing forest types, proximity of predators, and fruit tree dispersal. Therefore, travelling heights used for analysis were generalised based on a variety of sources. *C. cephus* is said to prefer lower levels of the canopy and understory, with most of their time spent between a height of 5 and 20m (Fleagle 1988). *C. nictitans* differs from *C. cephus* in that they rarely come to the ground, utilising mostly the middle and upper canopy at a height of 10 to 30m (Fleagle 1988). *P. badius* also rarely come to the ground (McGraw & Bshary 2002), spending nearly 80% of their time travelling at a height of 15 to 40m (McGraw 1998).

To simplify analysis in this preliminary study we were interested in overall frequencies of fractures in the collection, to ascertain patterns within each group of a given species, rather than solely numbers of fractures per individual. This meant that in individuals who had more than one fracture only the most severe fracture was used in order to standardise the analysis, although all fractures were recorded. Following Sokal and Rohlf's (1995) recommendation for analysing row by column tables, G – tests for independence were performed to determine whether or not differences in fracture frequencies between the species were statistically significant. All resulting values are compared to the critical values of the

chi-square distribution with the appropriate degrees of freedom.

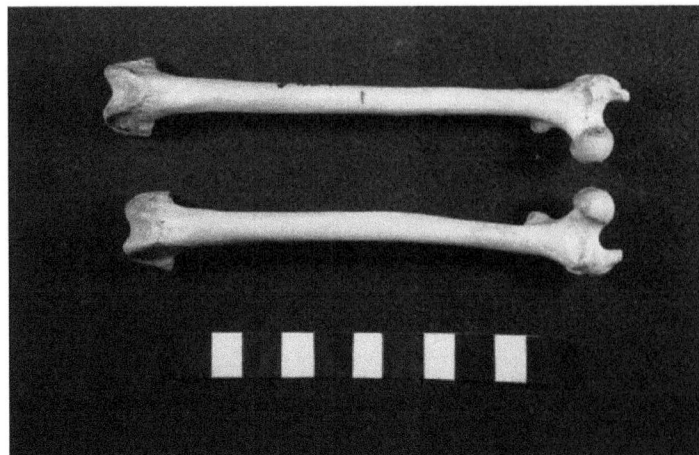

Figure 1. Fracture of left femur of C. cephus. *Scale divisions represent 1 cm.*

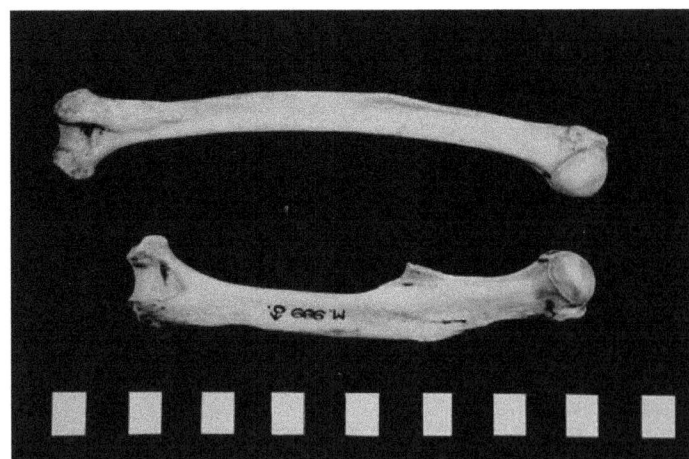

Figure 2. Fracture of the left humerus of C. nictitans.

Results

There were 15 primate skeletons with healed fractures out of the 80 examined. The majority of the specimens examined were adults (91% compared to only 9% of subadults). No fractures were observed among the subadults. Fracture frequencies ranged from 15% in *C. cephus* to 21% in *P. badius* (Table 2). Of the bones utilised in data analysis, the femur accounted for more than half of all fractures recorded (9 of 15; 60%), followed by the humerus (4 of 15; 27%) and radius (2 of 15; 13%). The observed frequencies of healed trauma were 21% for *P. badius*, 20% for *C. nictitans* and 15% for *C. cephus* (Table 2). Fracture frequency was found to be not significantly dependent upon travelling height.

There were no fracture complications, as determined by bone remodelling during and after healing, in *C. cephus* who travelled in the lower parts of the canopies.

TABLE 2. FRACTURE FREQUENCIES FOR EACH SPECIES EXAMINED.

Species	*n* with fracture	*n* without fracture	Fracture Frequency (%)
Cercopithecus cephus	4	23	14.8
Cercopithecus nictitans	5	20	20.0
Piliocolobus badius	6	22	21.4

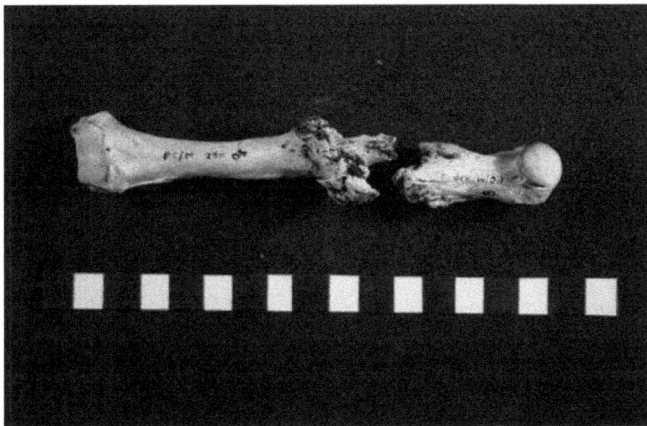

Figure 3. Left humerus of P. badius *with pseudoarthrosis and traumatic myositis ossificans.*

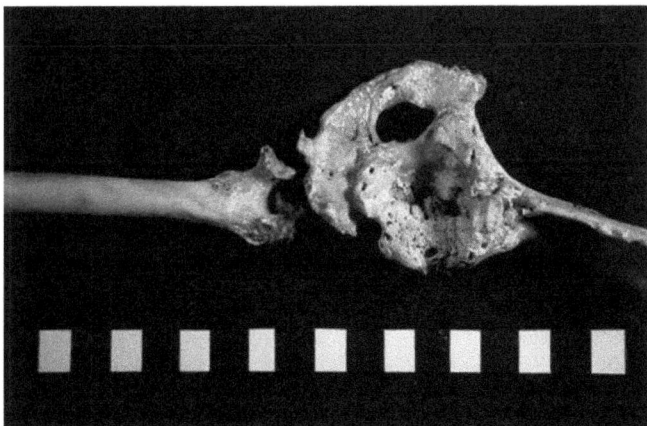

Figure 4. Left femur and associated pelvis of P. badius *with traumatic arthritis.*

However, in *P. badius* and *C. nictitans* several complications were observed. Of fractures among individuals in the mid and higher canopy, two of five (40%) and four of *six* (66.7%) showed evidence of complications in *C. nictitans* and *P. badius*, respectively. Fracture complications most often included traumatic arthritis and traumatic myositis ossificans (Figures 3 and 4). While there are marked differences between the species in fracture complications, once again they are not statistically significant.

The sex distribution of the skeletons overall was male (61%) and female (39%), however the distribution of male and female within each species was variable and this was dependent upon availability within the collection at the Powell Cotton Museum (Table 1). The difference of frequencies within the sexes is an issue that needs to be addressed in future studies of skeletal trauma to enable unbiased sex based analysis. However, from the availability of skeletal material examined at Powell Cotton there was found to be no statistical significance between fracture and sex

Discussion and Conclusion

Whilst fracture frequencies in the long bones were highest among *P. badius* and there seemed to be a trend toward increasing fracture frequency related to increased canopy height, it was not statistically significant. However, species utilising the higher canopy showed more instances of complications per fracture. Occurrence of fractures was also found to be independent of sex.

These findings do not initially support Preuschoft's (1990) assertion that the force acting on the body is greatly increased when falling from higher canopy heights. However, *P. badius, C. nictitans* and *C. cephus* are all arboreal quadrupeds using various portions of the tree canopy. *P. badius* and *C. nictitans* had the highest respective number of fractures, although *C. cephus* with a fracture frequency of 15%, while placed as the lowest canopy user (0 - 15m), utilises a range of canopy heights that sometimes overlap with heights of greater than 15m. It is difficult to ascertain at this time and without further analysis whether fractures are sustained while using the higher or lower canopy type within this particular species.

Comparison of these results with other studies is difficult because of the lack of similar published results comparing trauma to travelling height, as well as the lack of trauma studies on the three species examined here. As noted above, previous researchers have suggested that 20-30% of individuals within most primate groups should exhibit healed fractures (Lovell 1991; Schultz 1956a). The current study found healed fractures at an overall frequency of 18.75% Although slightly lower relative to these previous studies, the frequencies are not truly comparable as this study focused on the long bones only whereas other studies have looked at the complete skeletons.

While the findings of this preliminary work have shown that there is no relationship between trauma and travelling height in the three arboreal quadrupeds examined, a larger more detailed study is required which focuses on all other aspects of primate behaviour to determine possible reasons for differences or similarities in fracture frequencies. For better comparison of overall fracture frequencies to other studies, trauma should be included from all skeletal elements. Data should also be analysed with respect to each particular bone type; for example, comparison of all humeri or radii for each species. If breakage patterns emerge, they may provide insight into the processes involved. Additionally, information on time allocation among the various species in living samples would provide a better understanding of how much time and at which level of the canopy risky activities (potentially leading to trauma) are being performed. To summarize, this study is unique in examining trauma in three different contemporaneous species of arboreal quadrupedal primates who were collected from the same region and offers further information toward the characterization of trauma in non-human primates.

Acknowledgements

We would like to thank the Powell Cotton Museum in Birchington, Kent for access to their extensive primate skeletal collections and especially Malcolm Harmon for his wisdom regarding those collections. Thank you also to Dr. Nick Newton-Fisher for his insight and advice regarding primate behavioural observations.

References

Bramblett CA. 1967. Pathology of the Darajani baboon. *American Journal of Physical Anthropology* 26: 331-340.

Buikstra JA. 1975. Healed fractures in *Macaca mulatta*: age, sex and symmetry. *Folia Primatologica* 23: 140-148.

Duckworth WLH. 1911. On the natural repair of fractures, as seen in the skeletons of anthropoid apes. *Journal of Anatomy and Physiology* 46: 81-85

Fleagle JG. 1988. *Primate Adaptation and Evolution.* Academic Press, Inc: New York.

Jurmain R. 1989. Trauma, degenerative disease and other pathologies among the Gombe chimpanzees. *American Journal of Physical Anthropology* 80: 229-237.

Lovell NC. 1990. *Patterns of Injury and Illness in Great Apes: A Skeletal Analysis.* Smithsonian Institution Press: Washington, DC.

Lovell NC. 1991. An evolutionary framework for assessing illness and injury in nonhuman primates. *Yearbook of Physical Anthropology* 34: 117-155.

McGraw WS. 1998. Comparative locomotion and habitat use of six monkeys in the Tai Forest, Ivory Coast. *American Journal of Physical Anthropology* 105: 493–510.

McGraw WS and Bshary R. 2002. Association of terrestrial mangabeys (*Cercocebus atys*) with arboreal monkeys: Experimental evidence for the effects of reduced ground predator pressure on habitat use. *International Journal of Primatology* 23(2): 311-325.

Nakai M. 2003. Bone and joint disorders in wild Japanese macaques from Nagano Prefecture, Japan. *International Journal of Primatalogy* 24 (1): 179-195.

Ortner D. 2003. *Identification of Pathological Conditions in Human Skeletal Remains.* Academic Press: London.

Preuschoft H. 1990. Gravity in primates and its relation to body shape and locomotion. In *Gravity, Posture and Locomotion in Primates.* Jouffroy F K, Stack M H, and Niemitz C (eds.). *Il Sedicesimo.* Firenze; 109–127.

Sokal RR and Rohlf FJ. 1995. *Biometry.* Freeman and Company: New York

Schultz AH. 1937. Proportions, variability and asymmetries of the long bones of the limbs and clavicles in man and apes. *Human Biology* 9:281-328.

Schultz AH. 1956a. *The Life of Primates.* Weidenfeld and Nicolson: London.

Schultz AH. 1956b. The occurrence and frequency of pathological and teratological conditions and of twinning amongst non-human primates. *Primatologia* 1: 965-1014.

White TD. 2002. *Human Osteology.* Academic Press: San Diego.

Developmental stress and its morphological correlates

C. Buckley

Abstract

Growth and development in the human body is susceptible to environmental fluctuation. The environment has the potential to modify the individual during growth either through adaptation in form or state to change (phenotypic plasticity) or as a response to environmental stress (developmental instability). The impact of varying environmental conditions on cranial and facial morphology is examined in two London populations, dating from the 18th-19th century, from Old Church Street, Chelsea, and Redcross Way, Southwark. They represent socio-economically advantaged and disadvantaged social groups respectively and are predicted to show variation in cranial morphology as a result. Traits measured include cranial length, breadth and height, to indicate size and shape, and the widths of the orbits and first mandibular molar, to investigate fluctuating asymmetry. The Chelsea group were found to have a greater width for height than the Redcross group, contrary to predictions based on the results of previous studies, and a significantly larger circumference, which may indicate an enrichment of the environment. The Chelsea population also showed a more predictable cranial morphology than the deprived Redcross population. Finally, there was clear evidence of asymmetry in the Redcross orbits and molars, demonstrating a high degree of developmental instability in this group indicative of a poor environment of growth.

Keywords: Fluctuating asymmetry; developmental instability, growth, environment

Developmental plasticity and fluctuating asymmetry

Phenotypic plasticity affects both the actions of natural selection on phenotypes and the genetic response to selection (Stearns 1992), and can be defined as the 'ability of the individual to be adaptively modified by response to the environment during growth' (Roberts 1995). Plasticity allows an organism to react to its environment during growth in order to respond most efficiently to any changes. This response can take two forms, that of phenotypic plasticity or intra-individual variation, and developmental instability, a more extreme outcome to environmental stress. The latter will demonstrate irreversible changes to the adult phenotype and can be detected in skeletal remains, and may have a consequent impact on survival and reproduction (Schell 1995). The dichotomy between the response to variation (plasticity) and the response to stress (fluctuating asymmetries) is crucial to appreciating the nature of environmental impact. Plasticity is a long-term adaptation with an underlying genetic structure that allows for a controlled reaction to oscillations in environmental conditions. Fluctuating asymmetries arise in an individual's lifetime and are a marker of stress and instability during the period of growth.

The concepts of phenotypic plasticity and developmental instability have both been examined with regard to modern human populations. The impact of the environment on patterns of growth was first investigated by Boas in 1912 in the United States. He divided immigrants in American cities into 'types' and compared the features of those born abroad, those of the same type born in America within ten years after the arrival of the mother, and those born ten years or more after the arrival of the mother (Boas 1912). Concentrating on head form (expressed by the cephalic index, width taken as a percentage of the length), he found that in the American descendants of Central European types (Bohemians, Slovaks and Hungarians, and Poles), the length and width of the head both decrease. In his Hebrew group, the length of the cranium increased whilst width decreased, and the Sicilians and Neapolitans indicated an opposite trend. He believed that the influence of the American environment 'increases slowly with the increase of the time elapsed between the immigration of the parents and the birth of the child'; the foreign-born and American-born children of the same parents differed in cranial shape, with the latter group varying from their siblings and parents. These results suggested that something in the new environment had promoted an alternative pattern of growth. Furthermore, plasticity of development is not confined to the effects of migration; any environmental change such as food availability and quality, temperature, humidity, disease and activity, will have similar effects (Roberts 2001).

Boas's conclusions of the susceptibility of the skull to environmental change set a strong precedent for looking at cranial morphology in terms of developmental instability or environmental stress. Angel (1982) proposed that the cranial base is sensitive to nutritional status, as it must support the weight of the brain and head during growth, in the same way that the pelvic inlet, also affected by malnutrition, must support the weight of the body (Angel & Olney 1981). Failure of the skull to maintain its shape in this way, regardless of cause, results in platybasia, a flattening of the base of the skull. If this condition was to occur throughout a group, rather than varying individually through genetic influence, there is more likely to be an environmental, nutritional basis.

The skull heights of a skeletal population composed of healthy, mostly middle class Americans (countrywide, generally non-urban in origin) were compared with those of less nourished, socio-economically deprived individuals. This latter group comprised part of the Terry Collection, a large skeletal sample of low income

Americans. Angel measured porion-basion height and multiplied it by 100, then divided the result by biauricular breadth to control for gross skull size. The difference in skull height between the two populations was statistically significant when calculated in this manner, with an 11% increase in skull base height in the advantaged group (gross cranial size increase was much lower at 1%, as was stature increase at 3%). While the correlation between skull base height and pelvic inlet depth can in individuals be influenced by genetics, hormones and social practices, the relationship holds on a population level (Angel 1982).

Environmental variability may exceed the tolerance levels of a species and provoke short-term, non-genetic changes to the skeleton. Fluctuating asymmetry (FA) refers to random variation in bilateral traits and can act as a measure of developmental instability (as opposed to developmental plasticity), as development has to repeat itself twice, once on each side of the body (Hasson & Rossler 2002). Departures from FA include directional asymmetry (deviations from symmetry with a mean different from zero) and antisymmetry (deviations from symmetry with a bimodal distribution) (Black-Samuelsson & Andersson 2003). Developmental instability can be representative of the health of a population, since it is based on the assumption that environmentally or genetically induced deviations from the ideal phenotype are an indication of the precision of development, with lower precision (thus greater asymmetry) resulting from the disruptive effects of environmental stressors, poor genetic quality, or both (Kellner & Alford 2003).

Investigation of developmental plasticity and FA in the adults and juveniles of the two contrasting populations in this study will be restricted to the measurement of orbit height and width and various cranial dimensions. The interpretation of these results has implications for understanding the time scales involved and the means of adaptation employed. If, for example, FA is seen in various traits in adults, or in juveniles, but not in both, it may be a signal of a stressful environmental fluctuation that occurred only during the developmental period of that particular group. The response will then be plastic. However, if there is a difference in morphology at the population level the response is more likely to be genetic and the result of an adaptation to a change in environmental conditions.

A case study of growth and development in two London populations

The sample
The two populations used in this study are curated at the Museum of London and originate from 18th-19th century London, giving some control over the time period and geographical setting, thus providing a more viable comparison between the two. Selection was guided both by availability of large skeletal groups and the existence of an environmental dichotomy between those groups. The first population comprised the inhabitants of the tenement buildings in Redcross Way in Southwark

('Redcross'), and who were buried in Southwark poor persons' graveyard, St. Saviour's Burying Ground. The excavations of this site yielded a total of 166 burials, the majority of which are dated to the 19th century (Brickley 1999). This study uses a random selection of 78 individuals to give an approximate cross section of the population. The Redcross population is considered here as 'disadvantaged' since there is likely to have been malnourishment, and crowded and stressful living conditions for the people involved.

The second population was from Old Church Street, Chelsea ('Chelsea'), excavated under difficult conditions in 2000 (Cowie 2000). Known to be a more affluent population relative to that from Southwark (from the presence of lead-lined coffins, and burial on consecrated ground, contrary to Redcross), of the 288 individuals found, 70 were assessed for inclusion in this study. Again, selection was random. To provide an historical framework for each of these populations, and to ensure that the samples obtained are representative of the population of London as a whole, the demographic data from London's Bills of Mortality are used as a control, and indicate a juvenile mortality of 45-57.7% (Molleson & Cox 1993).

More juveniles appear to have died in the 'poor' Redcross sample than did in the 'rich' Chelsea sample. Of the total Redcross population, just under half (48.6%) were classed as juveniles, in marked contrast to the 10% juvenile occurrence in the Chelsea collection. However, this is not to say that the infant mortality rate in the Chelsea area was so low. It has been reported for the Spitalfields assemblage that despite an underlying juvenile mortality rate in London of 45-57.5%, the frequency of juveniles in the crypt reached only 18.9% (Molleson & Cox 1993: 208-9). While that of Chelsea remains somewhat lower, regardless of the random sampling employed, it seems likely that the principle reason accounting for such a difference between Redcross and Chelsea is potentially a class distinction in burial practices of the young. It is clear that the juvenile section of the Chelsea population, and perhaps also of Redcross, is under-represented.

In Redcross, the modal age at death is 25 years, with two smaller peaks at 30 and 40 followed by a fairly steep drop to approximately 60 years or older. The standard deviation is 11.07, indicating a relatively large degree of variation, and the mean age at death is 36.9 years. The Chelsea example has a mean age at death of 38.1 years and demonstrates a more gradual decrease after the modal ages of 25 and 35. However, the difference in mean age at death between these two populations is not significant ($p = 0.658$).

Predictions
It is possible to make some predictions regarding the morphological divergence between the advantaged and disadvantaged populations, based on what is already known about how certain features tend to behave in different environmental conditions:

1. Redcross should display greater variation in bilateral orbital symmetry than Chelsea, with wider confidence intervals from the increased developmental instability.

2. The width of left and right M1 should not differ significantly either within or between groups.

3. Redcross should have a flatter cranial base and be grouped above the regression line. Chelsea, with a greater skull base height, should be found below the line.

4. Chelsea should have a larger head, apparent both in a larger overall circumference and cranial capacity, and in a regression of length against breadth, where this population will group further up the line than Redcross. Relationships between variables will also be more stable in Chelsea.

Results – Fluctuating Asymmetry
The first prediction stated that the poor population would have a wider scatter on a regression of left orbit width against right, since there would be greater variation in the bilateral measurements as a result of developmental instability; while the second maintained that molar width would not differ appreciably between populations. Orbit width appears to show little difference between the two populations (Fig. 1).

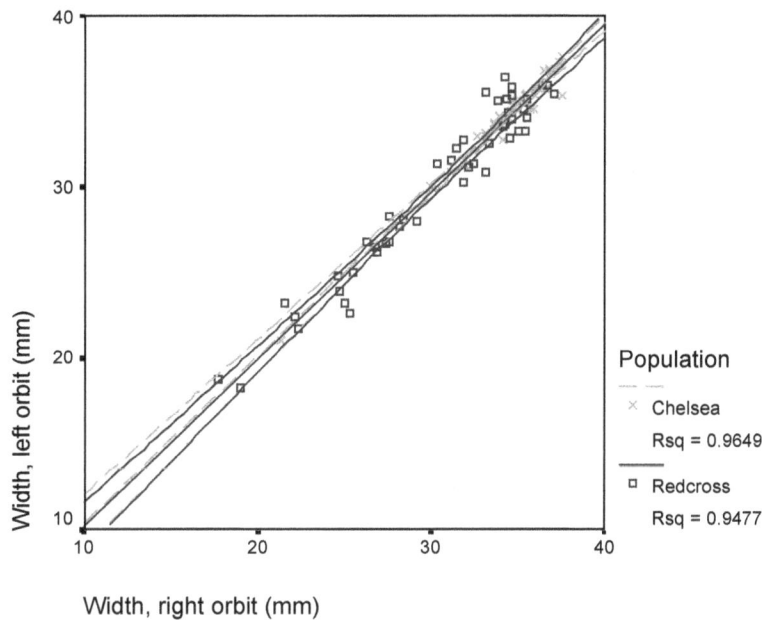

Figure 1. Left vs. right orbit width in the Redcross and Chelsea populations

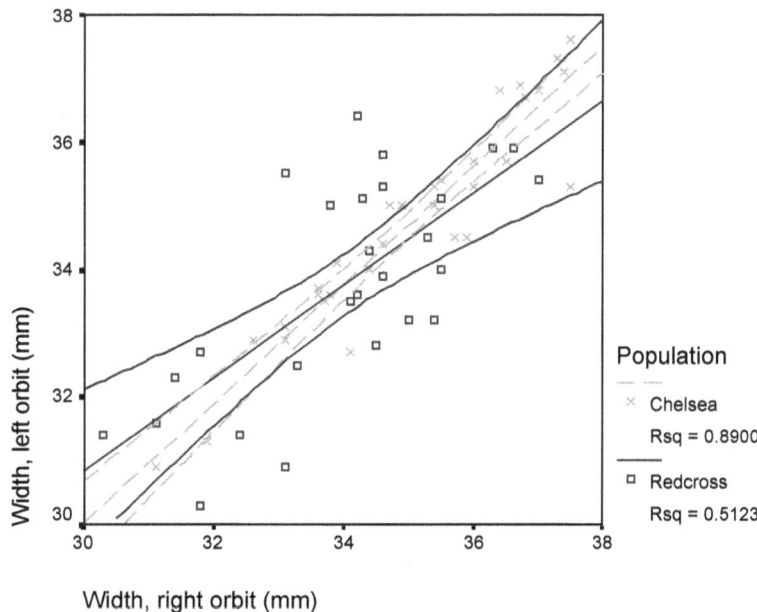

Figure 2. Left vs. right orbit width in the Redcross and Chelsea adult population

21

There is an excellent correlation between the two orbits, implying a high degree of symmetry and thus little developmental instability. However, when the juvenile results were removed from the regression, the disjunction between the two groups becomes more apparent. Despite the inequality in the proportion of juveniles in each population, their removal produced an almost balanced number of individuals in each group: 27 in Redcross, and 31 in Chelsea.

The removal of the juvenile measurements reduces the r-squared value of the Redcross line from 0.95 to 0.51, and the scatter of points is wide. Notably, when the juvenile measurements are regressed on their own they generate an r-squared value of 0.92, n = 20. Much of the Redcross correlation thus seems to be due to a relative lack of variation in the juveniles. To confirm this, the regression statistics show that the slopes of the lines produced by the juvenile regression are almost 1; in the Redcross group the value of the slope is 0.905, while for the Chelsea group it reaches 1.03. Moreover, the narrower confidence bands surrounding the Chelsea adult regression point to less developmental instability, as predicted earlier. The adult trend for the Redcross sample displays wider bands, and the increased presence of such instability in the adult population indicates that they may have experienced

harsher conditions than the next generation at the same developmental stage.

The directionality of orbit width, and the increase in the size of the Chelsea orbits over those of Redcross, is clarified when the $R_i - L_i$ values were plotted as a histogram (having first been divided by the mean to control for size), since fluctuating asymmetry is characterised by a normal distribution about a mean of zero (Kellner and Alford 2003).

There is a wide distribution of orbit width variation in the Redcross population (Fig. 3), with the largest peak occurring towards the right of the graph, and a smaller peak on the left to suggest fluctuating rather than directional asymmetry (although without a normal distribution)

The distribution for the Chelsea population (Fig. 4), by contrast, exhibits less asymmetry in orbit width in accordance with previous predictions. Here, the largest peak occurs at zero, indicating that the majority of individuals in this population were symmetrical in orbit width. Again, there is a tail towards the right, but to a far lesser extent than that seen in Redcross.

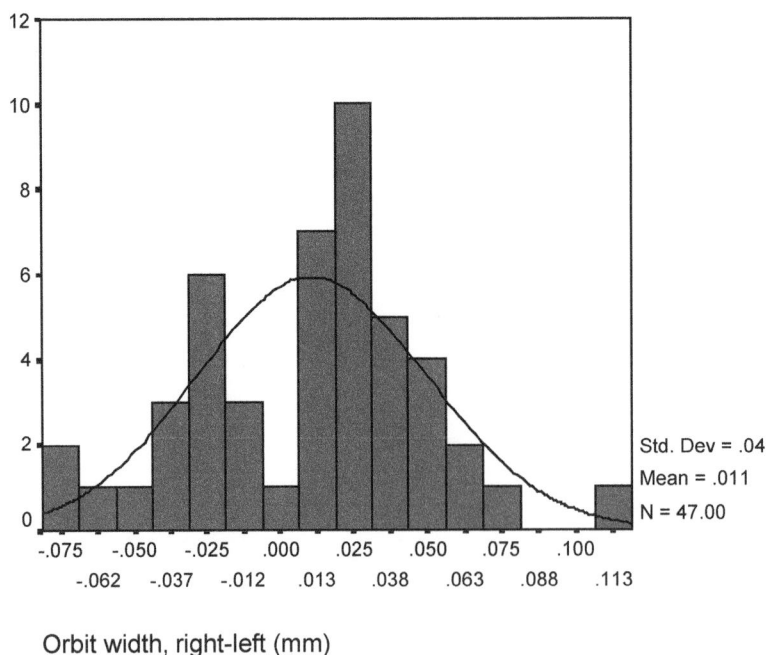

Figure 3. Distribution of R – L orbit width for the Redcross population

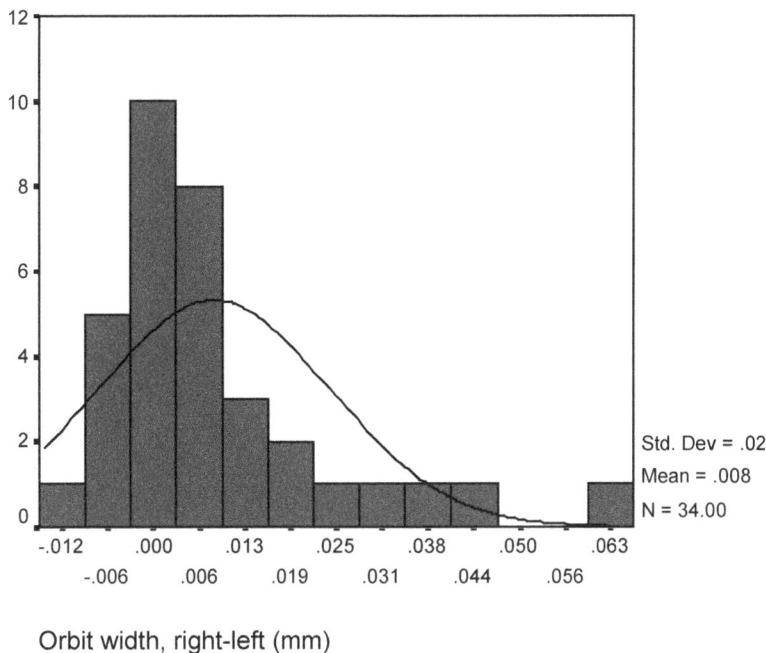

Orbit width, right-left (mm)

Figure 4. Distribution of R – L orbit width for the Chelsea population

Orbit width appears more subject to fluctuating instability than orbit height, presumably because of the constraints of the size of the eye itself. Fewer individuals retained enough of an intact facial structure to permit the measurement of this variable (12 in Redcross, 18 in Chelsea) but the result is convincing. The left and the right orbits registered a mean height of around 35mm for both populations. The standard deviation value was higher than that seen for orbit width, and the results of the t-test recorded a p-value of 0.502 for the height of the left orbits (equal variances assumed and 28 d.f. for both; $F = 0.1$ and $t = 0.681$), and 0.908 for the right ($F = 0.8$ and $t = 0.117$). The little variation there is between populations is not significant.

Molar width produced a similar pattern. As predicted above, the mean width of the left molar for each population was within 0.2mm of the mean width of the right molar. Surprisingly, and contrary to prediction 2 above, a paired-samples t-test signified that the widths of the left and right M1 were significantly different in Chelsea ($p = 0.035$; $t = -2.53$ with 8 d.f.). However, this requires a reverse reading of the results to understand how asymmetry is concluded. Figure 5 below illustrates a fine example of fluctuating asymmetry in the Redcross population, which may explain why the mean width of left M1 was not found to be significantly different from the mean width of right M1 ($p = 0.943$; $t = 0.073$ with 15 d.f.):

Furthermore, the mean widths of the Chelsea molars were, like the orbits, greater, in this case by 0.7mm, than those of the Redcross sample. This difference was significant, with a p-value of 0.023 for the left molar (equal variances assumed and 23 d.f. for both; $F = 0.013$ and $t = -2.429$), and 0.008 for the right ($F = 0.986$ and $t = -2.9$).

Results - Platybasia

The analyses carried out thus far have indicated a strong tendency towards developmental instability in the 'poor' population, conforming well to predictions of growth in stressful environments. Prediction 3 proposed that the poor population would have a flatter cranium than their advantaged counterparts (Angel 1982). Accordingly the two relevant measurements, basion-bregma and auriculare-bregma, were regressed against each other for the adult members of the two populations.

There is a distinct shift upwards by the Chelsea population, showing a greater distance of auriculare-bregma (width) for basion-bregma (height), and thus indicating, contrary to that expected, a comparatively flatter cranial base in this population rather than in Redcross. However the r-squared value for Redcross (0.493) is not convincing, and the wide confidence intervals make it impossible to judge whether this apparent shift is a demographic reality. Neither measurement supplied a significant result in a t-test between the populations, with $p = 0.119$ for basion-bregma (equal variances are not assumed; $F = 5.313$ with 42.648 d.f. and $t = 1.589$), and $p = 0.966$ for auriculare-bregma (equal variance are assumed; $F = 0.667$ with 51 d.f. and $t = -0.043$). These results do not fit with the clear indications of developmental instability in the orbits and the molars.

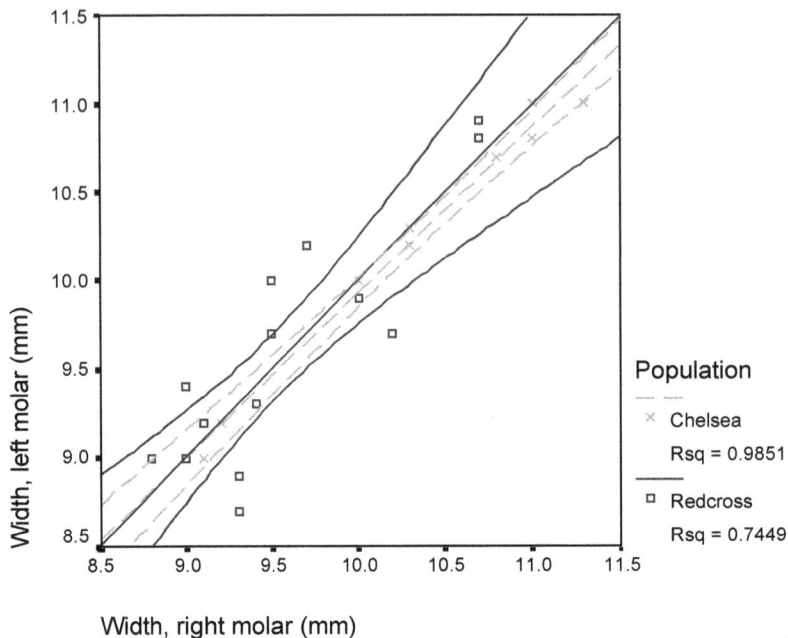

Figure 5. Left vs. right M_1 width in the Redcross and Chelsea adult populations

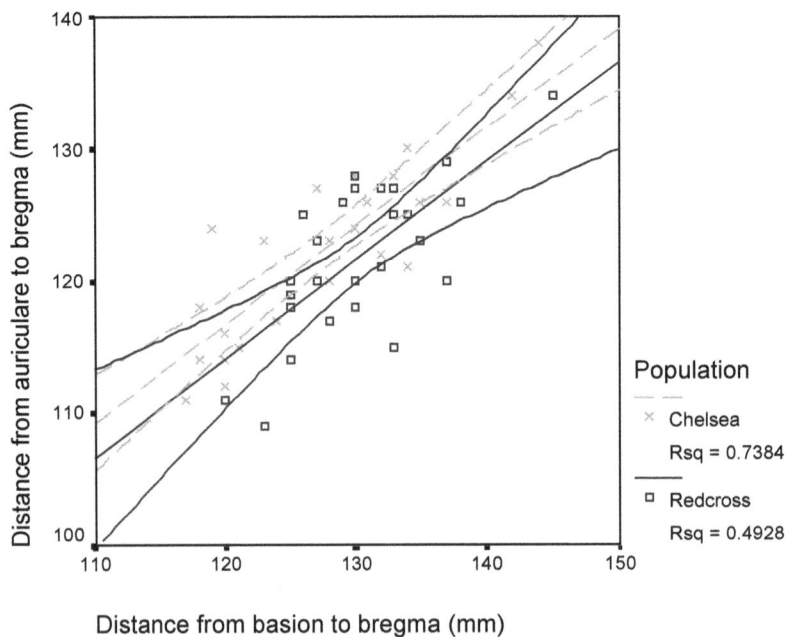

Figure 6 Assessment of adult cranial base height in the Redcross and Chelsea populations

Angel believed that platybasia was a symptom of environmental stress, caused by a weakening of the skull base with regards to the weight of the brain and head. He appeared to demonstrate this convincingly, but it has not been possible to replicate his results with these populations. On the contrary, Figure 5.22 suggests that the opposite is true; Chelsea appears to have a flatter cranial base. This may be explained through environmental sensitivity, with asymmetry a more refined measure of developmental stress than platybasia.

However, the problem may be with Angel's technique. It was mentioned above that although Angel controlled for width, he did not do so for circumference, and therefore gross overall cranial size. A *t*-test on circumference between the two populations in this study showed that the mean Chelsea circumference was significantly greater, with a *p*-value of 0.056 (forecast in prediction 4 above), although equal variances were not assumed ($F = 6.261$ with 44.701 d.f. and $t = -1.959$). The mean circumference of the Redcross sample was 517mm, while that of the Chelsea group reached 526mm and the low *p*-value can be accepted as demonstrating a significant difference.

To integrate this disparity, cranial height (basion-bregma) was divided by circumference and then regressed against bi-temporal width. Although these results are too widely distributed to be suitable for the fitting of a regression line or calculation of *r*-squared values (*r*-squared < 0.1 for both populations), there are nevertheless two distinct clusters. Redcross appears in the lower right corner while Chelsea is grouped higher up to the left, indicative of a larger circumference for height. This suggests that circumference has a strong enough impact through its relationship with cranial height to separate the two populations. This conclusion is further reinforced by a *t*-test carried out on this new variable between the populations, which shows a solidly significant difference in height controlled for circumference (*p* = 0.002. Equal variances assumed with 45 d.f.; $F = 3.176$ and $t = 3.353$).

These findings have implications for the acceptability of Angel's theory of platybasia. Figures 6 and 7 show that Chelsea, and not Redcross as Angel would have predicted, had the flattest skulls, although the difference between them was not significant. However, when height was controlled with circumference a significant result was generated, suggesting that Chelsea skulls were larger than those of Redcross in this latter dimension to a degree great enough to overrule the apparently flatter skulls. Angel's results have not been repeated here, and the question of platybasia remains unresolved, at least in these populations. The significant result seen in Figure 7 is due to circumference, and not cranial height.

Results – Cranial Variation
Circumference is increased in Chelsea, as noted above, but determining the origin of this increase is less straightforward. For bi-temporal width and maximum length, n = 28 for Redcross and n = 31 for Chelsea, and the mean difference between populations for these measurements is only 1-2mm. It is thus unsurprising that for length, *p* = 0.375 (equal variances are assumed; $F = 0.76$ with 57 d.f. and $t = -0.895$) and *p* = 0.211 for bi-temporal width (equal variances are not assumed; $F = 5.75$ with 52.398 d.f. and $t = -1.266$). A linear regression between these dimensions did not yield a discernable relationship; Chelsea produced an *r*-squared value of 0.33 in the adult population, while Redcross showed an even weaker trend with an *r*-squared value of 0.1.

A regression of circumference against width revealed a more apparent separation between the two adult populations. The stronger relationship, with *r*-squared = 0.78, was a product of the richer population compared to *r*-squared = 0.50 in Redcross. Figure 8 illustrates the discrepancy between the two samples, and indicates that for the Chelsea individuals, circumference is more influenced by an increase in cranial length than it is for the Redcross group.

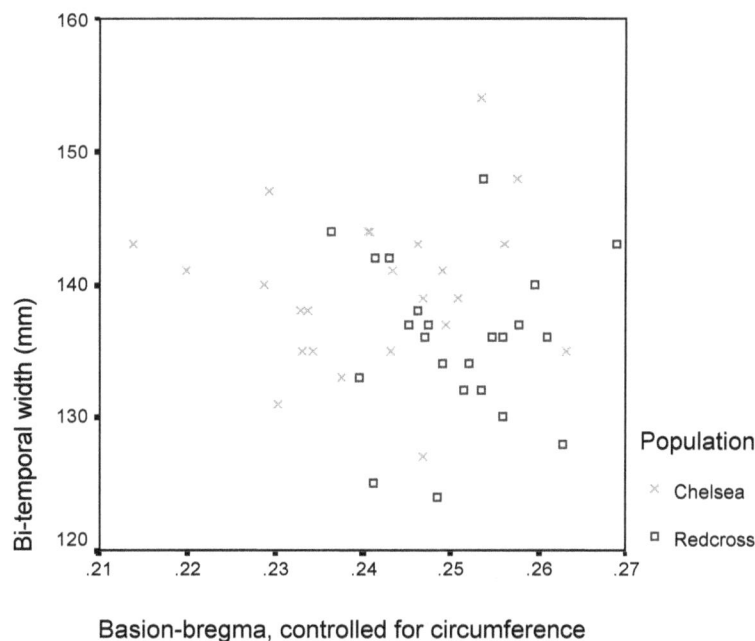

Figure 7 Relationship between cranial width and cranial height in the Redcross and Chelsea adult populations, with cranial height controlled for circumference

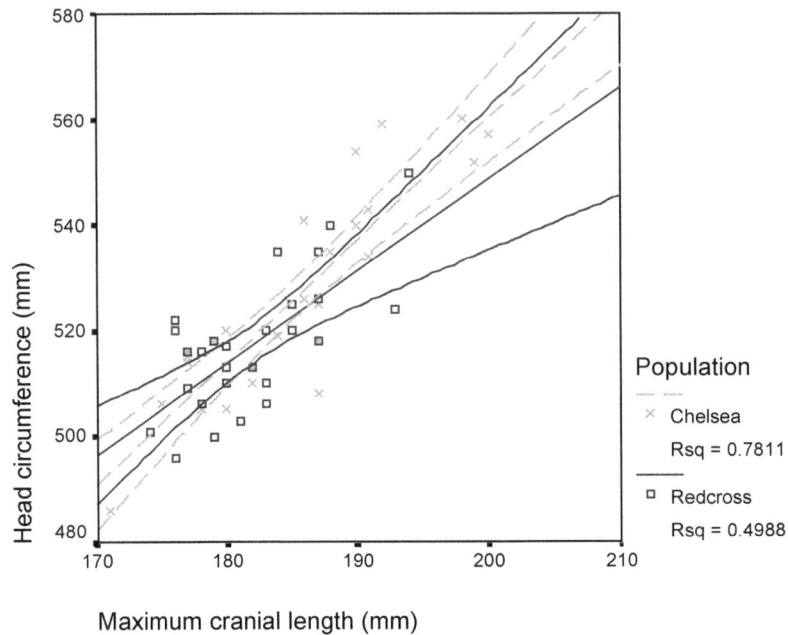

Figure 8. The relationship between circumference and maximum cranial length in the Redcross and Chelsea adult populations

A similar pattern is produced when bi-temporal width is regressed against circumference, the *r*-squared results almost matching those of Figure 8 above. With a much stronger trend in evidence for Chelsea, width is a better predictor of circumference for the rich group than for the poor.

Cranial height was regressed against cranial length and then bi-temporal width to produce a distinct shift upwards by Chelsea, implying that both length and width were greater for height in this population than in Redcross. However, in each case the *r*-squared value denoted a very weak relationship, similar to that seen in the platybasia investigations. The importance of these results lies in the inference that head shape is slightly different in the richer population. This difference is expressed most strongly through cranial circumference, although there is no increase in any one dimension that is solely responsible for this. Instead there is a mosaic of changes, linking length, breadth and height. Unfortunately, this is not reflected in cephalic index, which did not differ significantly between the two populations ($p = 0.921$ with 57 degrees of freedom) and so this study can find no support for Boas's findings.

Discussion
Developmental plasticity allows for adaptation to varying environments and thus improves the probability of survival. In more extreme conditions, developmental instability manifests as a permanent response to stress. Through the analysis of two modern human populations it has been possible to demonstrate the presence of developmental instability in an economically deprived group, signifying that there was a mismatch between the

Redcross population and its environment; or more accurately, that the environment did not fit the needs of its inhabitants. The lack of correlation between need and supply is seen through the directional and fluctuating asymmetry of the orbits, and through the variation in the relationship of various cranial measurements to each other.

The reasons for this variability are not clear, but this study suggests that cranial morphology is susceptible to environmentally-driven change, in the same way that orbit width is subject to developmental instability, albeit with less directional emphasis, and constrained by relatively greater canalisation. The presence of developmental instability has further repercussions for juvenile survival and the life history strategy of different populations, since the implication is for environmental stress and differential allocation of energy to growth.

Acknowledgements
My thanks are due to Bill White and Derek Seeley at the Museum of London for granting me access to the skeletons in their care, and for being so friendly and helpful. This work was undertaken as part of my PhD thesis at the University of Southampton, during which I was supported by the AHRB.

References

Angel JL. 1982. A new measure of growth efficiency: Skull base height. *American Journal of Physical Anthropology* 59: 297-305.

Black-Samuelsson S & Andersson S. 2003. The effect of nutrient stress on developmental instability in leaves of *Acer platanoides* (Aceraceae) and *Betula pendula* (Betulaceae). *American Journal of Botany* 90: 1107-1112.

Boas F. 1912. *Changes in bodily form of descendants of immigrants.* Columbia University Press: New York.

Brickley M. 1999. *The cross bones burial ground.* London: The Museum of London Archaeology Service.

Cowie R. 2000. *2-4 Old Church Street, Chelsea, London SW3: Progress Report (Unpublished).* Museum of London: London

Hasson O & Rossler Y. 2002. Character-specific homeostasis dominates fluctuating asymmetries in the medfly (Diptera: Tephritidae). *Florida Entomologist* 85: 73-82.

Kellner JR & Alford RA. 2003. The ontogeny of fluctuating asymmetry. *The American Naturalist* 161: 931-947.

Molleson TI & Cox M. 1993. *The Spitalfields Project volume 2: The anthropology. The middling sort. CBA Research Report 86.* Council for British Archaeology: York.

Roberts DF. 1995. The pervasiveness of plasticity. Mascie-Taylor CGN and Bogin B. (eds.) *Human variability and plasticity.* Cambridge University Press: Cambridge, 1-17.

Schell LM. 1995. Human biological adaptability with special emphasis on plasticity: History, development and problems for future research. Mascie-Taylor CGN and Bogin B (Eds.) *Human variability and plasticity.* Cambridge University Press: Cambridge, 213-237.

Stearns SC. 1992. *The evolution of life histories.* Oxford: Oxford University Press.

TEETH AND DIET: WHAT MORE IS THERE? TEETH AS MARKERS FOR POPULATION HISTORY

S.R. Zakrzewski

Abstract
Teeth have long been used to assess diet or disease status in populations but have also more recently been employed to assess population history and patterns of migration and movement. Recognition and definition of human osteological populations, and therefore migrations of those populations, within the archaeological record has often been problematic. Using an ancient Egyptian case study, this paper explores the potential of simple raw dental measurements for delineating population mobility. This approach is then used to assess the form of migration along the Nile Valley over the period of the development and intensification of agriculture and state formation.

Keywords: Teeth; population history; Egypt; mobility

Introduction
Tooth length and breadth constitute some of the most widely documented anthropometric data. Due to their high heritability (h^2), crown measurements such as these can be used to assess population or sample variation, such as by reference to classic 'odontographies' (Hillson 1996, 1-2). As teeth are particularly resistant to taphonomic and other destructive effects, they have potential within the study of human diversity. For one population and for a single sex, tooth size measurements should present with a normal distribution. Teeth show correlations in size with each other (summarised in Hillson 1996, 72-73) and yet exhibit distinct patterns of crown diameter variability (Kieser 1990). This paper presents a case study as an example to demonstrate the research potential of these sometimes under-used data.

Populations and Migrations
How can populations be recognised? Before considering mobility, the issue of recognition of groups must be considered. The above question may seem relatively simple, but the context in which the individuals are to be studied will require variation in the methods by which populations, and sub-groups within them, can be recognised. Can population groups be recognised by archaeological signatures (other than archaeological typologies)?

As previously discussed (Zakrzewski 2002), past populations have been separated through methods that have been determined in a uniformitarian manner from modern groups. These factors include features such as economy and subsistence (Kelly 1995), language (Foley 1997; Nichols 1992), religion and social belief systems (Geertz 1966), tool use and typology (Mellars 1992), and marriage patterns (Goodenough 1970; Gough 1959). By contrast, the vast majority of osteological studies that have attempted to define samples, groupings and populations have tended to undertake this for either palaeopathological or palaeoepidemiological reasons (summarised in Waldron 1994). Those studies that have considered differences between groups have generally done so in terms of differences between vastly geographically disparate populations (e.g. Howells 1973; Irish 1998; Irish & Turner 1990; Lahr 1996; Turner 1976; Turner & Swindler 1978).

Studies of dental variation have tended to concentrate upon the use of particular non-metric dental traits, usually employing the Arizona State University (ASU) Dental Anthropology Scoring System (Turner *et al.* 1991). Employing this method, populations have shown distinct groupings and clusters within global patterns of morphology. Population groupings have been recognised within African (Chamla 198; Harris & Rathbun 1989, 1991; Irish 1998; Irish & Turner 1990), European (Berry 1976; Frayer 1977; Owens 2003), American (Harris & Rathbun 1989; Irish & Turner 1987; Jacobson *et al.* 1977; Mahaney *et al.* 1990), Asian (Kaifu 1999; Turner, 1976, 1987, 1990) and Australasian (Birdsell 1993; Howells 1976; Stefan 1999) samples. This method is thus highly applicable when dental wear is minimal. Can a method be devised to assess dental variation and population affinities using raw measurements of tooth size that may be applicable for intra-population studies? The current paper assesses this possibility using a small Egyptian sample. This paper considers only temporal changes rather than geographical differences.

Previous North African Dental Studies
As summarised earlier, dental traits, such as non-metric features, as scored by methods such as the Arizona State University dental anthropology system (Turner *et al.* 1991) have been employed as methods of assessing global population variability (Irish 1998, 2000; Irish & Turner 1990; Scott & Turner 1997; Turner 1990; Turner & Hanihara 1977). Within North African material contradictory patterns have been seen (Calcagno 1986; Carlson & van Gerven 1977; Irish 1997, 1998, 2000). Overall there is a wide-scale trend for decreased dental size in the post-Pleistocene. This trend accelerates over the transition to the Holocene, after which it diminishes and sometimes reverses in historical times (Macchiarelli & Bondioli 1986). A temporal trend for reduction in tooth sizes within Nile Valley populations has been noted over long time periods (Calcagno 1986; Carlson & van Gerven 1977; Chamla 1980; Smith 1980), but short-term trends have not been studied. The size of Neolithic teeth from Egypt is generally smaller than in other North African populations (Chamla 1980). Egyptian maxillary teeth are smaller than those of all Neolithic populations from the Maghreb, Europe, and the Sahara, whilst only the European Neolithic population have smaller mandibular teeth. No change in tooth size has been found in later

Nubian samples, from the Meroitic period through to the Christian period (Calcagno 1986). The dental studies can be divided into two camps; one suggesting that the teeth provide evidence of population continuity from the Neolithic to the early historic period, as supported by Calcagno, Carlson and van Gerven, whilst the other camp, as supported by Irish and Turner, argue for the presence of discrete population changes.

North African, and especially Egyptian material, has been noted for its high degree of dental wear (Grilleto 1977, 1978, 1979; Ibrahim 1987; Hillson 1978, 1979). This high degree of wear tends to obliterate all dental crowns, thereby rending impossible the application of the ASU dental anthropology system. This study therefore proposes to employ direct methods of tooth size and dental wear to assess for population diversity and variability within a North African, and specifically Egyptian, context, in order to evaluate the potential of the use of these raw data for the assessment of human variability and population affinity.

Egypt as a Case Study
The ancient population of the Egyptian Nile Valley provides a classic case study to assess methods of defining temporally successive populations, and gauging their actual reality as separate morphological groups within the total Egyptian population. Egyptologists have traditionally considered the ancient Egyptian population to be relatively uniform and genetically closed (Kemp 1989). This view may be related to the development of the discipline, and is directly contradicted by the view of many anthropologists, and especially palaeoanthropologists, who note the Nile Valley as a corridor for migration through the Sahara and out of Africa (van Peer 1998).

The Egyptians themselves recognised differences between themselves and other groups. Some of the earliest evidence of this comes from the start of the Dynastic period (*c.* 3100 BC), when a depiction shows King Den (one of the first Pharaohs) smiting his eastern, and especially Asiatic, enemies. During the middle and later Dynastic periods, such as the Middle Kingdom (hereafter referred to as MK, *c.* 2000 BC), depictions are

found of Nubian mercenaries (e.g. stelae from Gebelein, such as Boston MFA 03.1848, specifically call the individual depicted 'Nehesy', i.e. the ancient Egyptian name for Nubians; see Kendall 1997), and Bedouin chieftains (e.g. the wall painting from the tomb of Khnumhotep II at Beni Hasan, Metropolitan Museum of Art 33.8.17). Some of the stelae, such as Leiden F 1938/1.6, suggest that, of these non-Egyptians, at least the Nubian mercenaries had married Egyptian women (Fischer, 1961) and adopted Egyptian behaviour patterns. These stelae suggest that the Nubian mercenaries of the period lived with, and were buried near, the Egyptian community that they served and that they were buried in an Egyptian manner, whilst still being depicted as Nubian, thus retaining their own ethnic identity.

Materials & Methods
The populations studied are summarised in Table 1. All samples were studied by the author and were selected for their relative cranial completeness and preservation, and for the documentation regarding their excavation and collection. Only adult skeletons were studied, with fusion of the sphenooccipital synchondrosis, eruption of the permanent third molar or complete fusion of postcranial epiphyses being used markers of skeletal maturity. The sample consists of 418 individuals, or 2468 teeth. Many of the crania did not have associated mandibles, and hence larger sample sizes were obtained for maxillary teeth. The populations studied are all from a relatively localised area in Upper and Middle Egypt.

Maximum bucco-lingual and maximum mesio-distal diameters were measured for each tooth in the dental arcade, following Buikstra & Ubelaker (1994, 61-62). Maximum values were taken rather than contact values in order to reduce intra-observer error, following Calcagno (1986, 350-51). The most reliable indicator of size is tooth breadth (bucco-lingual distance), as the mesio-distal distances are more affected by occlusal and interproximal wear, and hence only bucco-lingual diameters have been included in the analysis presented. Dental wear was scored following Molnar (1971) and Ibrahim (1987). This method takes into consideration the location of the tooth in the dental arch.

TABLE 1. SAMPLES INCLUDED IN CURRENT STUDY

Period	Sample	Collection	N ♂	N ♀
Badari	El-Badari	Duckworth	22	27
EPD	Abydos, El-Amrah & Gebelein	NHM & Marro	39	41
LPD	El-Amrah & Hierakonpolis	Duckworth & NHM	31	41
EDyn	Abydos & El-Amrah	Duckworth & NHM	55	42
OK	Regagnah, Meidum & Gizeh	NHM & Vienna	60	38
MK	Gebelein	Marro	13	9

Where Badari refers to the Badarian population (*c.* 4000 BC), EPD to the Early Predynastic (*c.* 3800-3400 BC), LPD to the Late Predynastic (*c.*3400-3100 BC), EDyn to the Early Dynastic (*c.* 3100-2700 BC), OK to the Old Kingdom (*c.* 2700-2400 BC) and MK to the Middle Kingdom (*c.* 2000 BC).
Duckworth refers to the Duckworth Collection of the University of Cambridge, NHM to the Natural History Museum, Marro to the Marro Collection of the University of Turin and Vienna to the Naturhistorisches Museum in Vienna.

Results

Employing paired t-tests (Sokal & Rohlf 1995, 356), no significant size differences were found in the bucco-lingual measurements of the right and left antimeres, and hence mean values were employed to maximise the sample size (number of individuals). Changes in tooth size variability through time were assessed using a type I hierarchical linear model (Sokal & Rohlf 1995, 272-317), correcting initially for sex (Table 2).

Only the lower second premolar (LP2) exhibits statistically significant change in bucco-lingual diameter through time. As this tooth exhibits no sexual dimorphism, Figure 1 plots this diameter across time with sexes pooled. This graph confirms the results of Scheffé's post-hoc test, which attributes the significance result to the LPD being significantly different in size from the EPD and MK samples; the LPD LP2 sample is, however, very small both in absolute terms and in comparison with the other periods.

TABLE 2. RESULTS OF UNIVARIATE ANOVA, TYPE I MODEL FOR B-L TOOTH DIAMETER.

Tooth	n	Sex		Period		Sex & Period interaction	
		F	p	F	p	F	p
UI1	74	2.99	0.089	0.47	0.795	0.29	0.884
UI2	108	8.94	**0.004**	0.60	0.702	0.26	0.936
UC	173	15.79	**<0.001**	2.11	0.067	1.37	0.237
UP1	234	8.31	**0.004**	1.40	0.224	2.35	**0.042**
UP2	239	2.03	0.156	1.27	0.280	1.78	0.119
UM1	266	6.94	**0.009**	0.44	0.820	3.53	**0.004**
UM2	258	11.19	**0.001**	1.85	0.103	4.12	**0.001**
UM3	190	4.27	**0.040**	1.85	0.105	1.22	0.303
LI1	71	3.91	0.053	0.09	0.986	2.35	0.064
LI2	93	1.54	0.219	0.40	0.808	0.85	0.496
LC	117	12.76	**0.001**	1.44	0.217	1.84	0.127
LP1	145	4.64	**0.033**	0.25	0.937	2.75	**0.031**
LP2	153	0.35	0.554	4.11	**0.002**	1.31	0.269
LM1	176	0.57	0.450	0.38	0.857	1.98	0.084
LM2	181	8.80	**0.003**	1.01	0.414	1.97	0.102
LM3	152	3.61	0.060	1.04	0.397	2.03	0.094

Significant results (at the 95% level) are shown in **bold**.

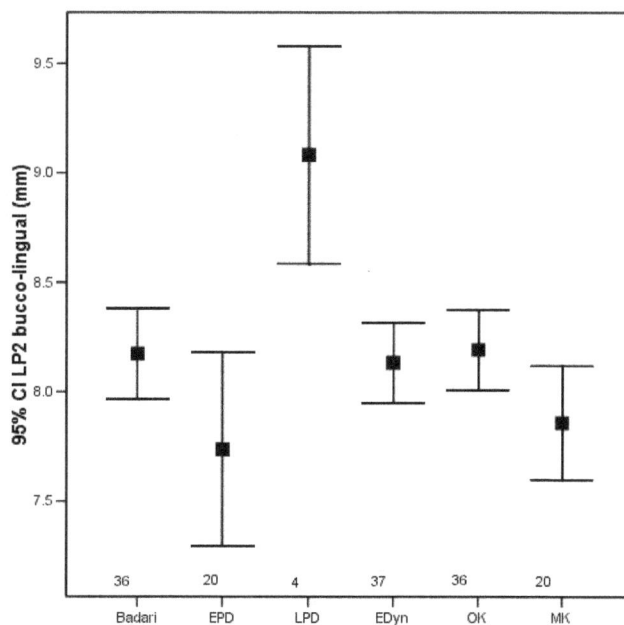

Figure 1. Change in bucco-lingual diameter of LP2 through time, sexes pooled. Sample sizes and 95% confidence intervals shown.

31

The significant interactions between sex and time period indicate that the male and female means for that tooth differ in pattern through time periods. In one case (LP1) this is the result of the females from the EDyn having relatively large measurements, and in two cases (UM1 & UM2) this is the result of the females in the MK having anomalously small teeth. The final case (UP1) both of these patterns occur. All these interactions are likely the result of having low female sample sizes for that tooth in one time period.

As shown in Table 2, 6 maxillary and 3 mandibular teeth exhibit significant sexual dimorphism in size. An example, UM3, is shown in Figure 2.

The Wilcoxon matched-pairs signed-ranks test (Sokal & Rohlf 1995, 440-444) found no left / right antimere

differences in dental wear for all teeth (except UC [$Z = -2.27$, $p < 0.05$]) and hence the mean wear code was employed in analysis. Dental wear is, by its very nature, an age-related phenomenon. This paper assesses the change in pattern of dental wear within the mouth across time periods. This was undertaken by computing an age-independent wear value, by dividing the mean wear code for each tooth by the mean wear value of the first molar. Employing the Mann-Whitney U test (Sokal & Rohlf 1995, 427-431) no significant sexual dimorphism in age-independent wear pattern was found.

Significant changes in wear pattern through the mouth were found between time periods (Table 3, Figure 3) by employing Kruskal-Wallis tests (Sokal & Rohlf 1995, 423-427).

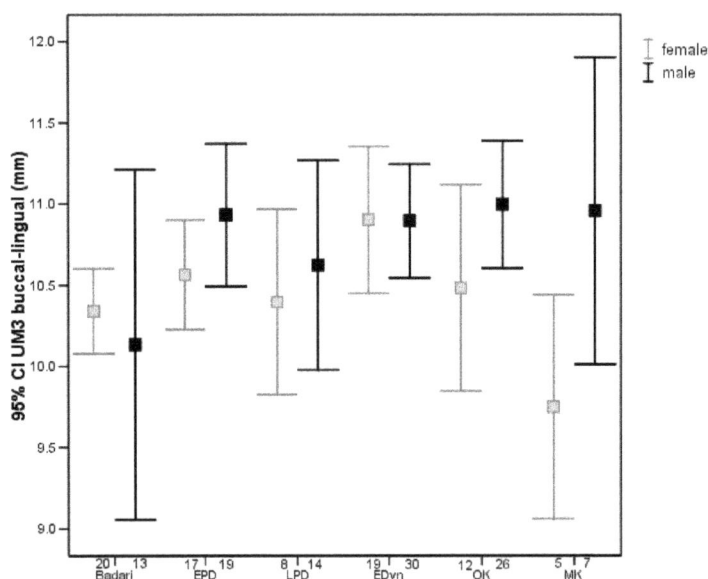

Figure 2. Change in bucco-lingual diameter of UM3 through time. Sample sizes and 95% confidence intervals shown.

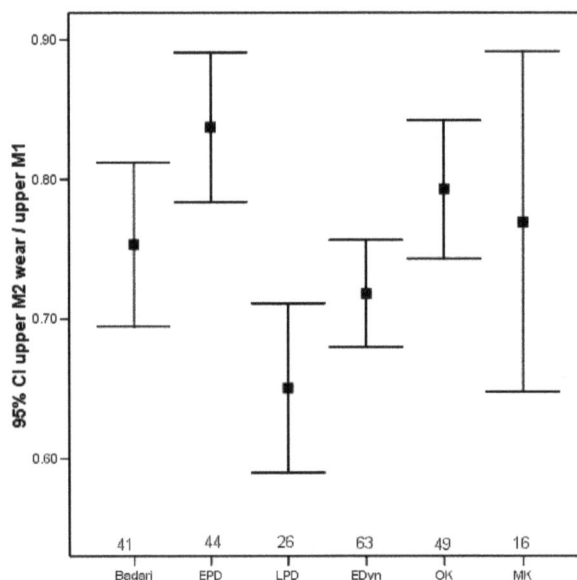

Figure 3. Change in dental wear in upper molars (M1 & M2) through time, sexes pooled. Sample sizes and 95% confidence intervals shown.

TABLE 3. RESULTS OF KRUSKAL-WALLIS TESTS, FOR POOLED SEX SAMPLES

Teeth	n	χ^2	p
UI1 / UM1	43	3.25	0.661
UI2 / UM1	57	6.17	0.290
UC / UM1	121	11.9	**0.036**
UP1 / UM1	204	14.4	**0.013**
UP2 / UM1	219	13.3	**0.021**
UM2 / UM1	239	25.6	**<0.001**
UM3 / UM1	172	21.4	**0.001**
LI1 / LM1	50	6.79	0.148
LI2 / LM1	64	7.58	0.108
LC / LM1	75	8.17	0.086
LP1 / LM1	124	13.0	**0.023**
LP2 / LM1	136	16.4	**0.006**
LM2 / LM1	163	3.04	0.694
LM3 / LM1	130	1.23	0.942
Significant results (at the 95% level) are shown in **bold**.			

Discussion

Significant sexual dimorphism has been demonstrated in tooth size measurements, but no real change has been demonstrated through time. The only tooth to exhibit a significant change through time is the lower second premolar, and this is likely an artefact of small sample size. Sexual dimorphism was found for tooth size for most teeth. It is also possible that the level of this sexual dimorphism increased through time.

Changes have been shown in dental wear patterning, with greater wear being present in the anterior portion of the arcade in the Dynastic period than in the Predynastic period, and more wear to the posterior dentition in the early Predynastic than in later periods.

Both these changes demonstrate that gross dental differences can be identified within temporally successive Egyptian populations. The actual morphological changes themselves, and the linked behavioural patterning, are hard to assess as they form a complex pattern of associated character traits. These changes could be associated either with population increase, increased gene flow through exogamy and migration or due to massive social changes associated with increasing social hierarchy and state formation.

The MK sample appears the most heterogeneous. This is in agreement with archaeological evidence from the stelae described earlier. Stelae found in the vicinity of Gebelein (the source of the MK sample in this study) indicate that from the First Intermediate Period (i.e. the period immediately preceding the MK), Gebelein had a colony of Nubian mercenaries. These stelae also suggest that these Nubians lived with and married Egyptian women (Fischer 1961). This would be predicted to increase the morphological heterogeneity of the immediately successive population, leading to a corresponding increase in both sexual dimorphism and in the biological variance of traits studied.

Conclusions

This study has shown that it is possible to use very basic dental data in order to identify morphological clusters within larger samples, and to develop hypotheses regarding population diversity and variability. The method is simple to apply but is effective in identifying the variability within the samples and thus to aid in understanding of the archaeological and social aspects of the population itself.

Acknowledgements
Many thanks are given to Rob Foley and Marta Lahr for their assistance with this project, and to the reviewer of this paper for helpful comments. Thanks are also given to the curators of skeletal material, for permitting me to study the skeletons in their care, particularly to M. Bellatti & M. Lahr (Duckworth, Cambridge), L. Humphrey & R. Kruszynski (NHM, London), E. Rabino-Massa & R. Boano (Marro, Turin) and M. Teschler-Nicola (NHM, Vienna). Funding was provided by the Wellcome Trust (Bioarchaeology Panel), Durham University (Addison Wheeler Fellowship) and University of Southampton.

References

Berry AC. 1976. The anthropological value of minor variants of the dental crown. *American Journal of Physical Anthropology* 45(2): 257-68.

Birdsell JB. 1993. *Microevolutionary Patterns in Aboriginal Australia*. Oxford University Press: Oxford.

Buikstra JE & Ubelaker DH. Eds. 1994. *Standards for Data Collection from Human Skeletal Remains*. Arkansas Archeological Survey Research Series. Fayetteville, Arkansas.

Calcagno JM. 1986. Odontometrics and Biological Continuity in the Meroitic, X-Group, and Christian Phases of Nubia. *Current Anthropology* 27(1): 66-69.

Carlson DS & van Gerven DP. 1977. Masticatory Function and Post-Pleistocene Evolution in Nubia. *American Journal of Physical Anthropology* 46: 495-506.

Chamla M-C. 1980. Étude des Variations Métriques des Couronnes Dentaires des Nord-Africains, de l'Épipaléolithique à l'Époque Actuelle. *L'Anthropologie* 84(2): 254-271.

Fischer HG. 1961. The Nubian Mercenaries of Gebelein during the First Intermediate Period. *Kush* 9: 44-80.

Foley WA. 1997. *Anthropological Linguistics: An introduction*. Blackwell: Oxford.

Frayer D. 1977. Metric Dental Change in the European Upper Paleolithic and Mesolithic. *American Journal of Physical Anthropology* 46: 109-120.

Geertz C. 1966. Religion as a cultural system. In Anthropological Approaches to the Study of Religion. Banton M. (ed.) *A.S.A. Monographs 3*. Tavistock: London; 1-46.

Goodenough WH. 1970. *Description and Comparison in Cultural Anthropology*. Lewis Henry Morgan Lectures. Aldine: Chicago.

Gough EK. 1959. The Nayars and the definition of marriage. *Journal of the Royal Anthropological Institute of Great Britain and Ireland* 89(1):23-34.

Grilleto RR. 1977. Carie et Usure Dentaire Chez les Égyptiens Prédynastiques et Dynastiques de la Collection de Turin (Italie). *L'Anthropologie* 81(3): 459-472.

Grilleto RR. 1978. Osservazioni sulla Carie Dentaria e sull'Usura dei Denti in una Serie di Crani Egiziani Predinastici. *Dental Cadmos* 46(1): 66-72.

Grilleto RR. 1979. Comparaison entre les Egyptiens Dynastiques d'Asiut et de Gebelen au Niveau de la Carie et de l'Usure des Dents. In *First International Congress of Egyptology - Actes*, 2-10 October 1976. Reineke WF. (ed.) Akademie Verlag: Berlin; 249-253.

Harris EF & Rathbun TA. 1989. Small tooth sizes in a nineteenth century South Carolina plantation slave series. *American Journal of Physical Anthropology* 78(3): 411-20.

Harris EF & Rathbun TA. 1991. Ethnic differences in the apportionment of tooth sizes. In *Advances in Dental Anthropology*. Kelley MA and Larsen CS. (eds.) Wiley-Liss: New York; 121-142.

Hillson SW. 1978. *Human Biological Variation in the Nile Valley, in Relation to Environmental Factors*. Unpublished PhD Thesis, Department of Anthropology. University College, London.

Hillson SW. 1979. Diet and Dental Disease. *World Archaeology* 11(2): 147-162.

Hillson SW. 1996. *Dental Anthropology*. Cambridge University Press: Cambridge.

Howells WW. 1973. Cranial Variation in Man. *Papers of the Peabody Museum of Archaeology and Ethnology*. Harvard University 67.

Howells WW. 1976. Physical variation and history in Melanesia and Australia. *American Journal of Physical Anthropology* 45(3 pt. 2): 641-9.

Ibrahim MA. 1987. *A Study of Dental Attrition and Diet in Some Ancient Egyptian Populations*. Unpublished PhD Thesis, Department of Anthropology. Durham University.

Irish JD. 1997. Characteristic high- and low-frequency dental traits in sub-Saharan African populations. *American Journal of Physical Anthropology* 102(4): 455-67.

Irish JD. 1998. Ancestral Dental Traits in Recent Sub-Saharan Africans and the Origins of Modern Humans. *Journal of Human Evolution* 34(81-98).

Irish JD. 2000. The Iberomaurusian Enigma: North African Progenitor or Dead End? *Journal of Human Evolution* 39: 393-410.

Irish JD. & Turner CG. 1987. More lingual surface attrition of the maxillary anterior teeth in American Indians: prehistoric Panamanians. *American Journal of Physical Anthropology* 73(2): 209-13.

Irish J & Turner C. 1990. West African Dental Affinity of Late Pleistocene Nubians. *Homo* 41(1): 42-53.

Jacobson A, Preston CB *et al*. 1977. The craniofacial pattern of the Lengua Indians of Paraguay. *American Journal of Physical Anthropology* 47(3): 467-72.

Kaifu Y. 1999. Changes in the Pattern of Tooth Wear from Prehistoric to Recent Periods in Japan. *American Journal of Physical Anthropology* 109(4): 485-499.

Kelly RL. 1995. *The Foraging Spectrum: Diversity in Hunter-Gatherer Lifeways*. Smithsonian Institution Press: Washington.

Kemp BJ. 1989. *Ancient Egypt: Anatomy of a Civilization*. Routledge: London.

Kendall T. 1997. *Kerma and the Kingdom of Kush 2500-1500 B.C.* National Museum of African Art, Smithsonian Institution: Washington.

Kieser JA. 1990. *Human Adult Odontometrics.* Cambridge University Press: Cambridge.

Lahr MM. 1996. *The Evolution of Modern Human Diversity.* Cambridge University Press: Cambridge.

Macchiarelli R. & Bondioli L. 1986. Post-Pleistocene Reductions in Human Dental Structure: A Reappraisal in Terms of Increasing Population Density. *Human Evolution* 1(5): 405-418.

Mahaney MC, Fujiwara TM *et al.* 1990. Dental agenesis in the Dariusleut Hutterite Brethren: comparisons to selected Caucasoid population surveys. *American Journal of Physical Anthropology* 82(2): 165-77.

Mellars P. 1992. Archaeology and the population-dispersal hypothesis of modern human origins in Europe. *Philosophical Transactions of the Royal Society of London* B 337:225-234.

Molnar S. 1971. Human Tooth Wear, Tooth Function and Cultural Variability. *American Journal of Physical Anthropology* 34: 175-190.

Nichols J. 1992. *Linguistic Diversity in Space and Time.* University of Chicago Press: Chicago.

Owens LS. 2003. Dental anthropology of the prehistoric Canarian islandscape: Tracking population dynamics and lifestyle. *American Journal of Physical Anthropology* S36: 162-163.

Scott GR & Turner CG. 1997. The *Anthropology of Modern Human Teeth: Dental Morphology and its Variation in Recent Human Populations.* Cambridge University Press: Cambridge.

Smith P. 1980. Regional Diversity in Epipaleolithic Populations. *Ossa* 6: 243-250.

Sokal RR. & Rohlf FJ. 1995. *Biometry.* W. H. Freeman: New York.

Stefan VH. 1999. Craniometric Variation and Homogeneity in Prehistoric /Protohistoric Rapa Nui (Easter Island) Regional Populations. *American Journal of Physical Anthropology* 110(4): 407-419.

Turner CG. 1976. Dental evidence on the origins of the Ainu and Japanese. *Science* 193: 911-913.

Turner CG. 1987. Late Pleistocene and Holocene population history of East Asia based on dental variation. *American Journal of Physical Anthropology* 73(3): 305-21.

Turner CG. 1990. Major features of Sundadonty and Sinodonty, including suggestions about East Asian microevolution, population history, and late Pleistocene relationships with Australian aboriginals. *American Journal of Physical Anthropology* 82(3): 295-317.

Turner CG & Hanihara K. 1977. Additional features of Ainu dentition. V. Peopling of the pacific. *American Journal of Physical Anthropology* 46(1): 13-24.

Turner CG & Swindler DR. 1978. The dentition of New Britain West Nakanai Melanesians. VIII. Peopling of the Pacific. *American Journal of Physical Anthropology* 49(3): 361-71.

Turner CG, Nichol CR *et al.* 1991. Scoring procedures for key morphological traits of the permanent dentition: the Arizona State University dental anthropology system. In *Advances in Dental Anthropology.* Kelley MA & Larsen CS. (eds.)Wiley-Liss: New York; 13-32.

van Peer P. 1998. The Nile Corridor and the Out-of-Africa Model: An Examination of the Archaeological Record. *Current Anthropology* 39(2): S115-S140.

Waldron T. 1994. *Counting the Dead: The Epidemiology of Skeletal Populations.* John Wiley: Chichester.

Zakrzewski SR. 2002. Exploring Migration and Population Boundaries in Ancient Egypt: A Craniometric Case Study. *Tempus* 7: 195-204.

Tracing change. Childhood diet at the Anglo-Saxon Blackgate Cemetery, Newcastle upon Tyne, England.

P.M. Macpherson, C.A. Chenery & A.T. Chamberlain

Abstract

There is little information about contemporary dietary practice of children in early medieval Europe. This paper presents results of isotopic studies on Anglo-Saxon childhood diet in the 8ᵗʰ to 11ᵗʰ century cemetery at Blackgate, Newcastle upon Tyne. Oxygen, carbon and nitrogen stable isotope measurements were made on dentine and enamel from 36 deciduous and permanent molar teeth from 8 adults and 6 juveniles. The results give indications of dietary habits from deciduous teeth at before birth to 8 months (m1 and m2) and from permanent teeth at 0-2.5 years (M1) 3.5-6.5 years (M2) and 9.3-12.6 years (M3). Adult values for carbon and nitrogen in bone collagen were measured in rib samples, age range: 15 to 39 years. Mean isotope differences were compared. Carbon and nitrogen results indicate a mixed terrestrial diet, possibly with some aquatic protein input. The composition of the diet changes throughout childhood, exhibited most strongly in the nitrogen results. Deciduous molars differ from permanent molars, indicating a transition phase towards lower $\delta^{15}N$ values between the age of 1 and 3 years. Between age 9 and 12 years the nitrogen isotope values converge on adult values as represented by rib bone collagen. The higher $\delta^{18}O$ in deciduous molars compared to permanent molars indicates a higher trophic level associated with breast feeding, which we estimate to cease before 1 year. The animal protein component of the diet decreases in early childhood, rising to adult levels as the child matures. We interpret these results as indicating that access to high quality food resources may be limited until a child is old enough to make a substantive contribution to the household economy.

Keywords: Childhood; Anglo-Saxon; diet; stable isotope analysis; bioapatite; collagen

Introduction

Food and diet are central to our understanding of societies past and present. Food is often seen in a number of different lights, as a source of nutrition, a social activity and a solitary vice, and there are many cultural dimensions to even the most mundane diet. While the biological necessity of eating and drinking remains unchanged over time, the cultural connotations vary. This is as true for children as for adults, even a biologically necessary process such as weaning has been shown through stable isotope analysis to vary through time and place (see for example Dupras *et al.* 2001; Fuller *et al.* 2003; Herring *et al.* 1998; White *et al.* 2004; Wright & Schwarcz 1999). There has been an implicit assumption that a child's diet after weaning converges on that of adults in the same population (Schurr 1998). This assumption is tested in the current research project through the use of stable isotope analysis.

A preliminary study into the longitudinal variation of childhood diet was carried out on individuals from the late Anglo-Saxon cemetery at Blackgate, Newcastle upon Tyne. Our aims were to determine what, if any, variation there was in an individual's diet over childhood and explore how this might be socially influenced. While the biology of the developing child plays a significant role in its food requirements, the attitudes and social influences of his/her immediate household and family, as well as the community at large, affect the type and quality of food made available for the child's consumption. A particular problem highlighted in previous isotopic studies of diet is the possibility that those children who died young may have had a different diet to those surviving to adulthood (Saunders & Hoppa 1993). This problematic issue was circumnavigated in this study by using the stable isotope ratios of carbon ($\delta^{13}C$), nitrogen ($\delta^{15}N$) and oxygen ($\delta^{18}O$) in permanent molar crowns from adult individuals to track dietary change; thus ensuring that individuals included in the study had survived their childhood and had at the least entered puberty and passed the Anglo-Saxon social age of majority of 12 years (Crawford 1999). Our methodology also provides information on the diet of individuals at discrete time intervals throughout childhood, allowing for the possibility of reconstructing individual dietary histories which could be compared with the information gained from their adult rib values. These data were supplemented by the analyses of deciduous molar crowns from juveniles to clarify the nature and timing of infant diet and the weaning process.

Documentary and archaeological evidence for the Anglo-Saxon period indicates a primarily agrarian society in which much food was produced locally. Cereal products such as bread and pottages, often including legumes, formed the backbone of Anglo-Saxon diet for all levels of society. This was supplemented by meat and dairy products for those with sufficient resources, with pork and game being seen as prestige meats. Fruits, nuts and vegetables were also eaten when seasonally available (Hagen 1992). Pearson (1997: 14) suggests that the foodstuffs available to the majority of Anglo-Saxon communities would have provided a "marginally adequate" diet according the standard United States Department of Agriculture food pyramid. Previous research has suggested that children may not have shared the same dietary habits as their parents; Crawford indicates that infants would have been breastfed for a prolonged period, until two or three years old, supplemented by soft white bread in the later stages (1999). After this period, Hagen has suggested that weaned children were not required to fast as intensely as their elders, and in monasteries were allowed greater proportions of animal protein.

The Blackgate Cemetery
The Blackgate cemetery is located in Newcastle upon Tyne, England and dates from between the late 7th and the early 12th centuries AD, the site is adjacent to the Norman castle. Post-conquest burials, probably relating to the castle, occurred in the cemetery, but all funerary activity ceased when a stone castle was built to replace the wooden keep in 1168. The cemetery was excavated between 1978 and 1992 and an assemblage of over 800 individuals was recovered. There is currently no settlement evidence associated with the cemetery, but it has been suggested that it may have been used by a lay community serving a monastery, or possibly a trading settlement (Merrony *et al.* 1996). The majority of the archaeological excavation was centred on the keep and underneath the railway arches of the nineteenth century viaduct. As only the northern extent of the cemetery could be defined, it is not possibly to determine the cemetery's original size. Most of graves were either plain or stone lined, but there is evidence for wooden coffins and cist burials in the cemetery, the latter are thought to date to the post Conquest period on stylistic evidence, and because many are cut into the castle fortifications. Individuals sampled for this study were taken from plain and stone lined graves.

Methods
Sampling protocol
Eight socially adult individuals (skeletal age range: 13 to 50 years) and six social juveniles (skeletal age range 1 to 10 years) were selected from suitable individuals from the Blackgate cemetery. Suitability was determined according to the following criteria: Individuals had a mandibular first, second and third permanent molar (for adults) or a mandibular first and second deciduous molar (for juveniles) of which at least 50% of the occlusal surface was present and the crown not destroyed by caries. Mandibular teeth were chosen as the age-related pattern of development is better characterised than for maxillary teeth (Hillson 1996). Samples were also taken from the rib of each adult, and control samples of faunal bone were included from sheep, cow, pig and dog remains recovered from the burial contexts.

Approximate biological ages of tooth formation are as follows: permanent first molar crowns form between 0 and 2.5 years; second molar crowns form between 3.5 and 6.5 years; third molar crowns form between 9.3 and 12.6years (Hillson 1996; Smith 1991). Deciduous first molar crowns form between 14.5 weeks after fertilization and 4 months after birth; second molar crowns form between 17 weeks after fertilization and 8 months after birth (Hillson 1996; Liversidge & Molleson 2004).

Application of carbon, nitrogen and oxygen isotope ratios in archaeological studies
Carbon and nitrogen stable isotope analysis of tooth dentine and rib was undertaken to determine the relative importance of animal and/or aquatic protein compared with cereal consumption during childhood from the onset of weaning to adolescence. Carbon stable isotopes are used in temperate zones such as Europe to determine the

input of terrestrial, freshwater and marine protein to diet. Humans eating a purely terrestrial diet of protein from animals who consume C_3 vegetation, will have a $\delta^{13}C$ value of -20 to -22‰. A marine input into the diet will cause those values to become less depleted (less negative) as the proportion of ^{13}C is higher in the marine carbonate reservoir and the effects of fractionation greater in the longer food chains; a diet consisting entirely of marine protein is likely to have a $\delta^{13}C$ values of -12 to -13‰ (Schulting 1988). The fractionation of carbon in freshwater systems is considerably more complex, but a similar, although less marked, enrichment is expected (Katzenberg & Weber 1999). The stable isotopes of nitrogen reflect protein intake and are principally used to place individuals within their local food web. $\delta^{15}N$ reflects the trophic level of the consumer, i.e., plant, herbivore and carnivore, with a 3‰ enrichment between levels, omnivorous consumers will display an intermediate value, reflecting the degree of carnivory in their diet. (Schwarcz & Schoeninger 1991). Due to the longer food chain, consumption of aquatic protein would produce further enrichment of $\delta^{15}N$ values. Analysis of faunal material commingled with the skeletal remains was undertaken to provide data concerning the local food web (Bocherens *et al.* 1991).

Oxygen stable isotope analysis was undertaken to determine weaning age, in conjunction with nitrogen isotopes and to provide a check for non-local individuals. Oxygen in bioapatite is derived from oxygen in body fluid, which in turn is derived from ingested water. There is a functional relationship between the isotopic composition of ingested water, body fluids and bioapatite controlled by metabolic fractionation (Levinson *et al.* 1987). Empirical equations, such as that of Levinson, which take into account metabolic fractionation, can be used to estimate the mean oxygen isotope value of an individual's drinking water. Wong *et.al.* (1987) demonstrated that the oxygen isotope ratio of breast milk is enriched in relation to ingested water; an elevated $\delta^{18}O$ ratio is therefore expected in teeth formed before weaning.

Analytical methodology
The entire crown surface of the tooth was abraded to a depth of > 100microns and the enamel and crown dentine were separated by drilling using a tungsten carbide dental bur. Enamel bioapatite was converted to silver phosphate following a method modified from Crowson *et al.* (1991) and O'Neil *et al.* (1994). Collagen was extracted following a modified Longin (1971) procedure (Brown *et al.* 1988) from tooth dentine and rib samples.

Analytical measurement of $\delta^{18}O$ was by high temperature conversion elemental analyser coupled to a continuous flow isotope ratio mass spectrometer. Analytical measurement of $\delta^{13}C$ and $\delta^{15}N$ were by an elemental analyser coupled to a continuous flow isotope ratio mass spectrometer.

Results
The average ratios for each sample type (animal, tooth type and rib) for carbon, nitrogen and oxygen stable

isotopes are given in Table 1. Carbon and nitrogen elemental ratios (C:N ratio) for human tooth and bone ranged between 3.0 and 3.7, faunal ratios ranged between 2.7 and 4.7; the acceptable range is between 2.9 and 3.6 (van Klinken 1999) indicating a much poorer level of preservation for the faunal bone. The relationship between carbon and nitrogen isotope values presented in figure 1, show the food web for the Blackgate humans and animals. All human values plot consistently higher

than the herbivore values derived from sheep and cow, with some pigs plotting an intermediate value and some plotting in the herbivore range. While pigs are traditionally considered to be omnivorous, the practice of pannage may account for the more herbivorous signal of some animals. The one dog sample plots in the same range as the humans, indicating a similar diet. The results thus demonstrating the principle of trophic isotope ratio shift with diet.

TABLE 1. STABLE ISOTOPE VALUES OF OXYGEN, CARBON AND NITROGEN FOR HUMAN AND FAUNAL SAMPLES FROM THE BLACKGATE CEMETERY. A LOWER CASE 'M' INDICATES A DECIDUOUS MOLAR, UPPER CASE, PERMANENT. SUB SAMPLES CONSIST OF ONLY THOSE SAMPLES IN A CATEGORY WITH ACCEPTABLE C:N RATIOS.

Sample type	Number	Mean $\delta^{13}C$	St Dev $\delta^{13}C$	Mean $\delta^{15}N$	St Dev $\delta^{15}N$	Mean C:N	St Dev C:N	Mean $\delta^{18}O$	St Dev $\delta^{18}O$
All cow	5	-21.97	0.4	6.30	0.7	3.5	0.8		
Subsample	1	-21.60		5.19		3.4			
All sheep	5	-21.94	0.1	6.98	1.7	3.3	0.4		
Subsample	3	-21.97	0.1	6.95	1.6	3.3	0.4		
All pig	6	-21.33	0.5	8.29	1.7	3.4	0.8		
Subsample	3	-21.21	0.5	7.74	2.0	3.0	0.1		
Dog	1	-19.80		11.27		3.0			
All m_1	6	-20.39	0.8	13.12	1.2	3.5	0.1	18.47	0.6
Subsample	5	-20.18	0.7	13.42	1.1	3.4	0.1		
All m_2	6	-20.50	0.7	12.76	1.0	3.4	0.1	18.44	0.5
All M_1	6	-20.62	0.5	11.98	2.1	3.4	0.1	17.75	0.6
All M_2	6	-20.64	0.6	10.69	1.3	3.5	0.2	17.58	0.4
Subsample	5	-20.54	0.7	10.44	1.5	3.4	0.1		
All M_3	6	-20.30	0.6	11.18	1.0	3.4	0.1	17.49	0.6
All rib	6	-20.34	0.4	10.88	0.3	3.2	0.2		

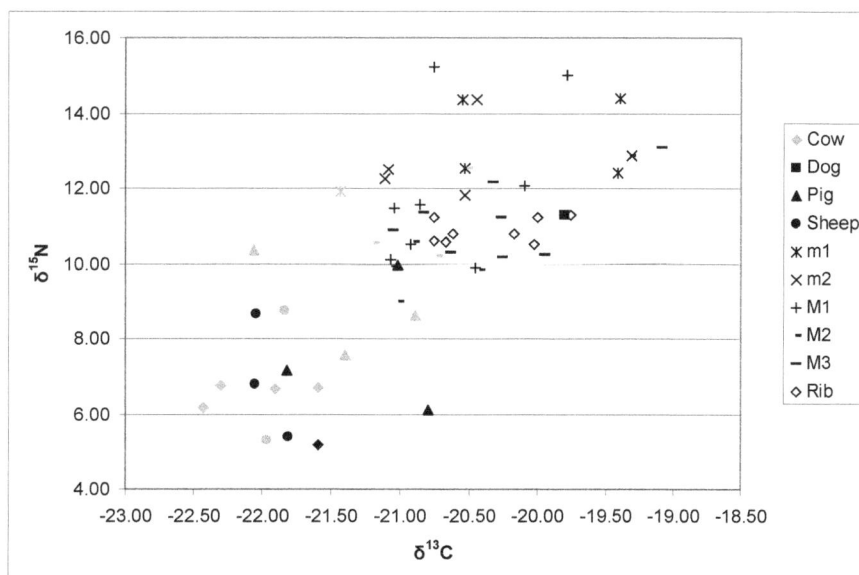

Figure 1: $\delta^{13}C$ and $\delta^{15}N$ values for all samples, points marked in grey have poor C:N ratios.

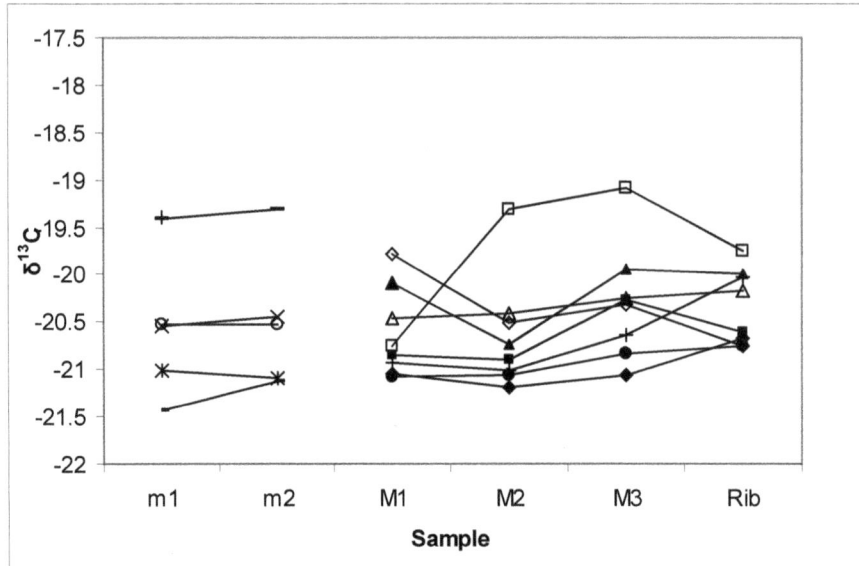

Figure 2: $\delta^{13}C$ values from individuals from the Blackgate Cemetery (lines indicate multiple samples from the same individual)

TABLE 2: RESULTS FOR STUDENT'S PAIRED T-TEST FOR DIFFERENCES BETWEEN TOOTH TYPES PRESENTED FOR EACH ISOTOPE

Pair	$\delta^{13}C$ significance	$\delta^{15}N$ significance	$\delta^{18}O$ significance
m1 – m2	0.139	0.545	0.205
m1 – M1	0.716	0.644	0.035
m1 – M2	0.615	0.019	0.019
m1 – M3	0.983	0.032	0.009
m1 – rib	0.762	0.010	-
m2 – M1	0.190	0.610	0.000
m2 – M2	0.685	0.033	0.004
m2 – M3	0.877	0.051	0.003
m2 – rib	0.758	0.024	-
M1 – M2	0.666	0.004	0.038
M1 – M3	0.206	0.134	0.073
M1 – rib	0.014	0.456	-
M2 – M3	0.026	0.048	0.933
M2 – rib	0.060	0.297	-
M3 – rib	0.204	0.873	-

Excluding values with poor C:N ratios, average herbivore $\delta^{13}C$ is -21.8‰, while pig is insignificantly different at -21.3‰. Carbon values for the Blackgate human skeletons range between -19‰ and -21.5‰ (Figure 2), indicating little or no marine animal protein in the diet. Student's paired t tests show that there are statistically significant differences between the carbon values of the M1s and ribs, and M2s and M3s at the 95% confidence level (see Table 2). There is therefore some fluctuation in carbon isotope intake over childhood. Ribs values are clustered between -19.7‰ and -20.8‰ indicating a relative homogeneity in protein sources of dietary carbon in adult life.

Average $\delta^{15}N$ for herbivores (again excluding those samples with poor C:N ratios) is 6.51‰ and for pigs is 7.74‰, indicating their more omnivorous diet. These faunal nitrogen isotope results show that the human adult rib values are positioned at least one trophic level above that of the herbivores, (trophic shifts from herbivore to carnivore are accepted as being +3 to +4‰ (Ambrose 1991). The consumption of pig protein, and possibly freshwater fish as part of the diet is therefore likely in this community. Average $\delta^{15}N$ values decline from 13.1‰ in the first deciduous molar to 10.9‰ in the ribs (Figure 3). Again, the adult rib values show the least amount of variation, the standard deviation for all 6 ribs being 0.35. Student's paired t-tests show that deciduous molars are significantly different from M2s, M3s and ribs. M1s are different from M2s and M2s are different from M3s. There is no significant difference between deciduous molars and M1s (table 2). These changes are likely to reflect the changes in protein intake throughout childhood: nitrogen isotope levels decrease on the advent of weaning (Balasse & Tresset 2002; Dupras et al. 2001; Herring et al. 1998; Katzenberg & Pfeiffer 1995; Richards et al. 2002) and continue to drop in early childhood, increasing towards adult levels as the child matures.

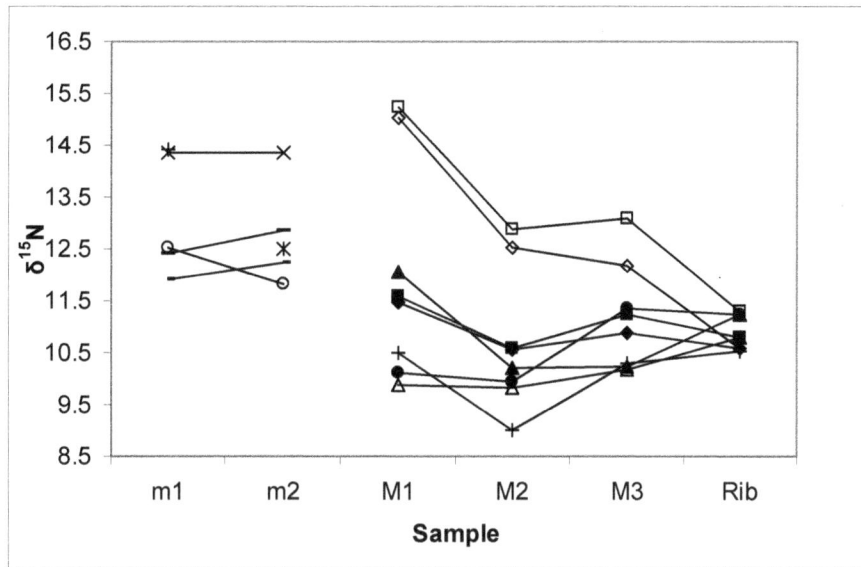

Figure 3. $\delta^{15}N$ values from individuals from the Blackgate Cemetery (lines indicate multiple samples from the same individual)

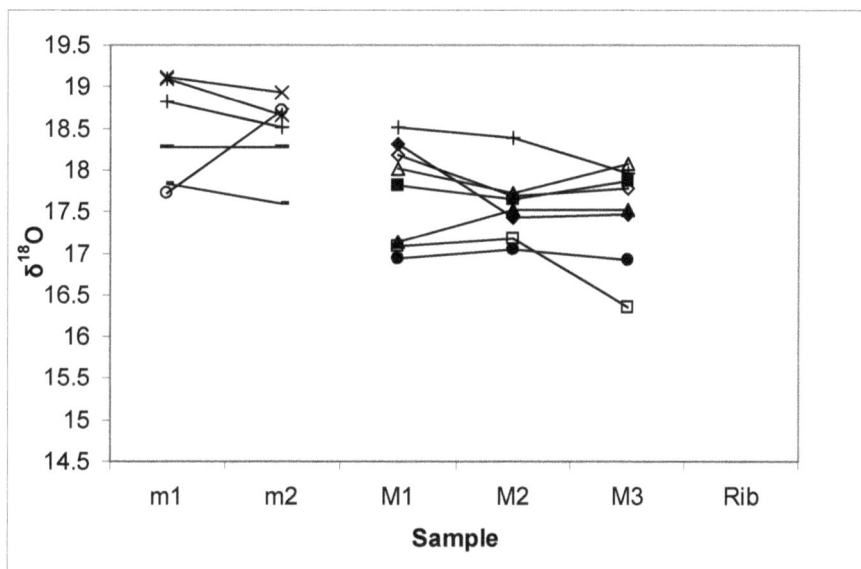

Figure 4. $\delta^{18}O$ values from individuals from the Blackgate Cemetery (lines indicate multiple samples from the same individual)

The oxygen incorporated into tooth enamel is derived primarily from ingested water. It is expected that tooth enamel, formed over a number of years, will contain an oxygen isotope signal reflecting average oxygen isotope values of the ground water of the place in which they resided during tooth formation. Blackgate oxygen results, do not uniformly match drinking water values expected for the area (Darling & Talbot 2003a, b), this study currently uses oxygen values solely to track the trophic shift associated with breast feeding (White *et al.* 2004; Wright & Schwarcz 1998, 1999) whilst an explanation is sought. The results show significant differences at the 95% confidence level between deciduous molars and all

permanent molars (Figure 4, table 2). This illustrates the shift between breastfed and weaned individuals with higher δ18O in the teeth formed during breastfeeding. There are also significant differences between the M1s and M2s at the 95% confidence interval and the M1s and M3s at the 90% confidence interval (Table 2), indicating that part of the M1 formed during breastfeeding, and that weaning occurred on average between 6 and 9 months.

Interpretation
These data show a consistent pattern of dietary change throughout childhood which contrasts with the relatively homogenous adult dietary signature. The oxygen results

41

indicate that, as expected, almost all the children in the sample received their nutrients through breast milk for at least the first six to nine months of their lives. Although a little less clear in places, the nitrogen isotope results corroborate this trend but continue to exhibit a fluctuating pattern throughout childhood. This time frame for the cessation of breastfeeding and introduction of solid food is in contradiction with Crawford's (1999) interpretations of the archaeological and documentary evidence for a late age of weaning. The drop in both oxygen, and to a lesser extent nitrogen values in the M1's suggest weaning occurred in the first year of life, possibly supplemented by a protein rich weaning food, for example a milky gruel. Ongoing analysis of oxygen and nitrogen isotopes in deciduous second molar roots (formed between approximately 10 months and 3 years) will help to further clarify this interpretation.

Nitrogen isotope values are depleted uniformly in permanent 2nd molars, in most cases to below adult rib levels. This suggests a much lower animal protein intake between 3 and 6 years compared to that of adults, possibly with more reliance on herbivore rather than pig protein. Values rise in later childhood to converge on adult isotope levels. While the carbon values are more variable, there is a tendency for them to become depleted during the formation of the M2, perhaps indicating that cereal and vegetable sources of protein have become more important in the diet. We interpret this to mean that these children are beginning a transition towards an adult diet during the formation of the third molar crown (between 9 and 12 years of age), perhaps as they commence more substantial contributions to the household economy.

The differentiated 'childhood diet' of the younger children offers two possible interpretations. The first is that in the post weaning period children were seen as needing a diet differently constituted to the adult members of their society, perhaps to aid growth and development; they were therefore given specialised food. Alternately, these children were seen as liminal members of the adult society, perhaps because they were not yet able to make a full contribution to the household, and were therefore not eligible to consume a fully adult diet inclusive of prestige items such as pork.

There are also two individuals who show very different carbon and nitrogen signatures from the rest of the sample. They display elevated nitrogen isotope ratios and much greater shifts in carbon values. It is possible that there was an aquatic element to the childhood diet of these two individuals. This probably derives from freshwater sources as carbon values are not enriched enough to indicate a significant marine input into the diet (Schwarcz & Schoeninger 1991).

Conclusion

This study demonstrates the applicability of stable isotope studies in the investigation of variation in childhood diet at the individual and population level. Diet amongst later Anglo-Saxon children varied in the amount of animal protein they ingested at different stages of their childhood, the lowest levels being eaten by young children (3 to 6 years). As children matured, the proportion of animal protein in the diet increased, and converged on adult values. There are two possible explanations for this:

- There was lowered parental and social investment in children until they were likely to survive childhood and make significant contributions to the domestic economy.
- Younger children were given a special culturally defined 'childhood diet' until they were deemed old enough to eat the same food as adults.

These competing 'nurture' and 'neglect' hypotheses are being investigated in more detail by further isotopic analyses both from the Blackgate material, and also from other sites.

Acknowledgements
Analytical support was provided by NERC Isotope Geosciences Facilities Steering Committee. Thanks to Dr Pia Nystrom (University of Sheffield) for facilitating access to the skeletal samples, Dr Paul Halstead for faunal identifications, and to Dr Dawn Hadley for discussions on Anglo-Saxon society. The University of Sheffield provided a PhD studentship and travel support for Miss Macpherson.

References

Ambrose SH. 1991. Effects of diet, climate and physiology on nitrogen isotope abundances in terrestrial foodwebs. *Journal of Archaeological Science* 18: 293-317.

Balasse M & Tresset A. 2002. Early weaning of neolithic domestic cattle (Bercy, France) revealed by intra-tooth variation in nitrogen isotope ratios. *Journal of Archaeological Science* 29: 853-859.

Bocherens H, Fizet M, Mariotti A, Lange-Badre B, Vandermeersch B, Borel JP, Bellon G. 1991. Isotopic biogeochemistry (13C, 15N) of fossil vertebrate collagen: application to the study of a past food web including Neandertal man. *Journal of Human Evolution* 20: 481-492.

Brown TA, Nelson DE, Vogel JS, Southon JR. 1988. Improved collagen extraction by modified Longin method. *Radiocarbon* 30(2): 171-177.

Crawford S. 1999. *Childhood in Anglo-Saxon England*. Stroud, Sutton Publishing.

Crowson RA, Showers WJ, Wright EK, Hoering TC. 1991. Preparation of phosphate samples for oxygen isotope analysis. *Analytical Chemistry* 63: 2397-2400.

Darling WG & Talbot JC. 2003a. The O & H stable isotopic composition of fresh waters in the British Isles. 1. Rainfall. *Hydrology and Earth System Sciences* 7(2): 163-181.

Darling WG & Talbot JC. 2003b. The O & H stable isotopic composition of fresh waters in the British

Isles. 2. Surface waters and groundwater. *Hydrology and Earth System Sciences* 7(2): 183-195.

Dupras TL, Schwarcz HP, Fairgrieve SI. 2001. Infant feeding and weaning practices in Roman Egypt. *American Journal of Physical Anthropology* 115: 204-212.

Fuller BT, Richards MP, Mays S. 2003. Stable carbon and nitrogen isotope variations in tooth dentine serial sections from Wharram Percy. *Journal of Archaeological Science* 30: 1673-84.

Hagen A. 1992. *A Handbook of Anglo-Saxon Food: Processing and Consumption.* Pinner, Anglo-Saxon Books.

Herring DA, Saunders SR, and Katzenberg MA. 1998. Investigating the weaning process in past populations. *American Journal of Physical Anthropology* 105: 425-439.

Hillson S. 1996. *Dental Anthropology.* Cambridge, Cambridge University Press.

Katzenberg MA and Pfeiffer S. 1995. Nitrogen isotope evidence for weaning age in a nineteenth century Canadian skeletal sample. *Bodies of Evidence. Reconstructing History Through Skeletal Analysis.* A. L. Grauer. New York, Wiley-Liss: 221-235.

Katzenberg MA & Weber A. 1999. Stable Isotope Ecology and Palaeodiet in the Lake Baikal Region of Siberia. *Journal of Archaeological Science* 26: 651-659.

Levinson AA, Luz B, Kolodny Y. 1987. Variations in oxygen isotopic compositions of human teeth and urinary stones. *Applied Geochemistry* 2(367-371).

Liversidge HM & Molleson TI. 2004. Variation in crown and root formation and eruption of human deciduous teeth. *American Journal of Physical Anthropology* 123: 172-180.

Longin R. 1971. New method of collagen extraction for radiocarbon dating. *Nature* 230: 241-2.

Mcrrony C, Boulter S, Rega E. 1996. *Male migration into medieval Monkchester: mobility and social role.* University of Sheffield Sheffield, Unpublished Manuscript.

O'Neil JR, Roe LJ, Reinhard E, Blake RE. 1994. A rapid and precise method of oxygen isotope analysis of biogenic phosphate. *Israel Journal of Earth Science* 43: 302-212.

Pearson KL. 1997. Nutrition and the early-medieval diet. *Speculum* 72(1): 1-32.

Richards MP, Mays S, Fuller BT. 2002. Stable carbon and nitrogen isotope values of bone and teeth reflect weaning age at the medieval Wharram Percy site, Yorkshire, UK. *American Journal of Physical Anthropology* 119: 205-210.

Saunders SR & Hoppa RD. 1993. Growth deficit in survivors and non-survivors: biological mortality bias in subadult skeletal samples. *Yearbook of Physical Anthropology* 36: 127-151.

Schulting RJ (ed) 1988. Slighting the sea: stable isotope evidence for the transition to farming in northwestern Europe. *Neolithic Studies.* Zalozila, Filozofska Fakulteta Oddelek Za Arheologijo.

Schurr MR. 1998. Using stable nitrogen isotopes to study weaning behaviour in past populations. *World Archaeology* 30(2): 327-342.

Schwarcz HP & Schoeninger MJ. 1991. Stable isotope analysis in human nutritional ecology. *Yearbook of Physical Anthropology* 34: 283-321.

Smith BH. 1991. Standards of human tooth formation and dental age assessment. *Advances in Dental Anthropology.* M. A. Kelly and C. L. Larsen. New York, Wiley-Liss: 143-168.

van Klinken GJ. 1999. Bone collagen quality indicators for palaeodietary and radiocarbon measurements. *Journal of Archaeological Science* 26: 687-695.

White C, Longstaff FJ, Law KR. 2004. Exploring the effects of environment, physiology and diet on oxygen isotope ratios in ancient Nubian bones and teeth. *Journal of Archaeological Science* 31: 233-250.

Wong WW, Lee LS, Klein PD. 1987. Deuterium and oxygen-18 measurement on microlitre samples of urine, plasma, saliva and human milk. *American Journal of Clinical Nutrition* 45: 905-913.

Wright LE & Schwarcz HP. 1998. Stable carbon and oxygen isotopes in human tooth enamel: identifying breastfeeding and weaning in prehistory. *American Journal of Physical Anthropology* 106: 1-18.

Wright LE & Schwarcz HP. 1999. Correspondence between stable carbon, oxygen and nitrogen isotopes in human tooth enamel and dentine: infant diets at Kaminaljuyu. *Journal of Archaeological Science* 26: 1159-1170.

GROWTH IN MODERN WESTERN CHILDREN: A REPRESENTATIVE SAMPLE?

Clegg M

Abstract

Most studies of growth and development in either early hominids or in archaeological populations rely on standards based on modern western children. This assumes that earlier populations follow the same growth trajectories as seen in present day westerners. However, non-western populations vary across a number of growth indicators. Data are examined from a number of published studies for both dental and skeletal development. Populations from USA, United Kingdom, India, Thailand, Japan, Canada, Bangladesh, Mexico, South Africa and Egypt are included. The results illustrate the wide range of variation in both tooth emergence and development exhibited within and between populations. Children may be much younger or older than their stage of dental development suggests. Contrary to expectations, the mean age for skeletal stage appears to be closer to the known age, this is particularly so for older children. This paper examines the implications for earlier populations to produce more realistic age ranges.

Keywords: growth; development; dental age; skeletal age; juvenile

Introduction

When skeletal remains are discovered one of the first questions is how old the individual was when they died. Researchers in this field strive to be as accurate as possible in determining age at death. The latest techniques are used and age is often recorded to the nearest month. We must, however, be realistic in our attempts to estimate age at death, and acknowledge that that there are limits to the accuracy achievable, even within living populations. To this end, the present study aims to show the range of variation that exists within modern populations of different ethnic origin and to apply these data to an archaeological sample.

Determination of age at death in children is based on a series of maturity indicators; these often include dental development, epiphyseal closure and long bone length or estimate of stature. These indicators are regarded as having varying degrees of reliability. Dental development is widely regarded as the most accurate (Smith 1991; Eveleth & Tanner 1990) and stature estimates the least (Eveleth & Tanner 1990; Tanner 1986). When assessing archaeological samples the mean age for stage of development is generally based on published developmental standards. These standards are generally from western populations of European origin; many growth tables use middle class American children. The mean development of these children with access to good health care and a nourishing diet may not be representative of many modern human populations (Lampl & Johnston 1996; Liversidge & Speechley 2001; Clegg & Aiello 1999). The difference between these modern populations and archaeological populations may be even greater (Clegg & Aiello 1999; Molleson & Cox 1993).

The stage of development or physiological age of an individual does not necessarily equate to chronological age (Tanner 1986; Eveleth & Tanner 1990; Hoppa & Fitzgerald 1999). The variation within a population may be wide, with children of the same chronological age being at different stages of physiological development. The age for dental stage in the healthy children used by Demirjian (1976, 1986) may vary greatly; for example a child from this population assessed as being 10 years old may in fact be between 8.8 yrs and 12.0 years of age (Demirjian 1976, 1986). The same holds true for skeletal age with children having a skeletal maturation age of 6 years old possibly having a chronological age between 4 years and 8 years of age (Jones et al. 1973). This is even more pronounced when growth standards developed for one population are transferred to another. Lampl and Johnston (1996) compared the skeletal and dental ages of Mexican children of the same chronological age and found that there was an error of up to 2 years under or over estimation of age for skeletal maturation, while dental development would only predict that children at that chronological age would be within a 3.5 year range. The children may look younger or older than their chronological age depending upon the local maturation rates (Tompkins 1996). These problems all compound the difficulties for the archaeologist or anthropologist in determining the age at death in an archaeological population. We have no idea of prevailing ecological conditions or local maturation rates.

This study will consider intra and inter specific variation in populations from different ethnic groups and from different regions of the world. The variation in rates of both dental development and epiphyseal closure will be examined. These data will then be applied to an archaeological sample of known age.

Method

The modern human sample comprises data from the ethnic groups detailed in table 1 taken from the published literature. These groups include data from, USA, United Kingdom, India, Thailand, Japan, Canada, Bangladesh, Mexico, South Africa and Egypt. The data for some ethnic groups are only available for dental development or skeletal maturation. Only the following groups have data for both indicators: British, Indian and Mexican.

TABLE 1: SOURCES OF DENTAL & SKELETAL DATA

Nationality	Source
British	Liversidge & Streetley 2001; Eveleth & Tanner 1976
Bangladeshi	Liversidge & Streetley 2001
Mexican	Lampl & Johnston 1996; Eveleth & Tanner 1990
Norwegian	Nykanen et al. 1998; Eveleth & Tanner 1990
Thailand	Eveleth & Tanner 1976
Japanese	Eveleth & Tanner 1976, 1990
India	Eveleth & Tanner 1990
USA	WHO website. Eveleth & Tanner 1990
Egyptian	Eveleth & Tanner 1976, 1990

The known age archaeological sample is composed of the children from Christ Church, Spitalfields London. These children are of Huguenot ancestry and were interred in the crypt of Christ Church during the 18th and 19th century. They are currently housed at the Natural History Museum London. These data are already published as part of a previous study (Clegg & Aiello 1999).

Two measures of dental development were compared to show both the intra and inter specific variation within and between groups. These methods include Demirjian's four or seven tooth method (1976, 1986) and the age of tooth emergence (Garn et al. 1967) will be compared to the US data as this is the standard currently used by the WHO (Garn 1965; Al-Mazrou et al. 2003).

Demirjian et al.'s method is used to estimate age by scoring the stage of tooth development for a selection of teeth. Tooth formation is divided into eight stages and criteria for the stages are given for each tooth separately. Each stage of the seven or four teeth is given score according to a statistical model, which has also been used for assessment of skeletal maturity (Tanner et al. 1975) standards are given for each sex separately . The scores for the seven or four teeth are then summed and this summed score is compared to the 50th percentile score to produce an estimated dental age (Demirjian 1976, 1982).

Tooth emergence is often considered to be the time at which the tooth breaks through the gum or the presence of the tooth at examination. It is generally considered less accurate then the stage of tooth development as the actual emergence is a fleeting event and the event may already have occurred prior to the examination.

Skeletal development is more problematic as direct comparisons with the archaeological sample are not possible. Skeletal age in living children is usually assessed by means of an atlas of skeletal maturation of the wrist bones (e.g. Tanner et al. 1975; 1983). The TW2 method examines 20 bones of the hand and wrist and the assignment of a letter score to each bone depending on

the attainment of clearly described bone-specific maturity indicators. This score is then converted to a numeric score using tables given by Tanner et al. (1983) and the scores are summed for each individual to give a maturity score on a scale of 0 to 1000. Maturity scores can be converted to skeletal age by comparing this score to British standards given by Tanner et al. (1983). The method commonly used for archaeological remains assesses the stage of fusion of all available bones (Krogman & Iscan, 1986). The skeletal age of the bones is inferred on the basis of epiphyseal closure using the five-point scale developed by McKern and Stewart on the basis of North American populations of European descent (in Krogman & Iscan 1986). However, as both standards have been developed using modern western children it is possible to show how closely these skeletal age estimates approximate chronological age.

Results

Dental development
Figure 1 shows the US population mean for stage of tooth emergence. The other studies on tooth emergence used here are taken from Eveleth and Tanner's book (1976) no range or standard deviation is quoted, and the original studies are difficult to obtain as they date from the 1950s and 1960s. The mean data for tooth emergence for the Thai, African, Japanese and Egyptian children (Figure 1) shows quite clearly that some populations are not near the mean for the American children. Thai children for example have a mean emergence age for upper premolar 2 and upper molar 1 that is older than the American mean emergence age. Other populations, for example the Japanese children always have the mean age of tooth emergence at a younger age than their American counterparts. Even within a population different teeth may emerge earlier or later than the reference mean. The Thai children as discussed earlier have an older emergence age for some teeth, while others, for example the upper second molar, have a younger mean emergence age. The Thai children could be assessed either younger or older on average than the American children depending on which teeth were available for analysis. This has implications for archaeological samples.

One method emerging as a developmental standard for tooth development is that of Demirjian et al. This method has now been tested in several populations worldwide. Figure 2 illustrates the deviation from chronological age that this method produces when applied to other populations. This figure also shows the range of variation within the French Canadian reference sample. The difference from actual chronological age is relatively small in the younger children, for example children aged under eight years old the difference between the age estimate and known chronological age is under one year in the British and Bangladeshi children, but far greater in the Southern Indian children, who may have a difference with chronological age as great as 2.64. The older children have an increasingly greater mismatch between known and estimated ages. The teenagers in the sample have a mismatch greater than four years. When these

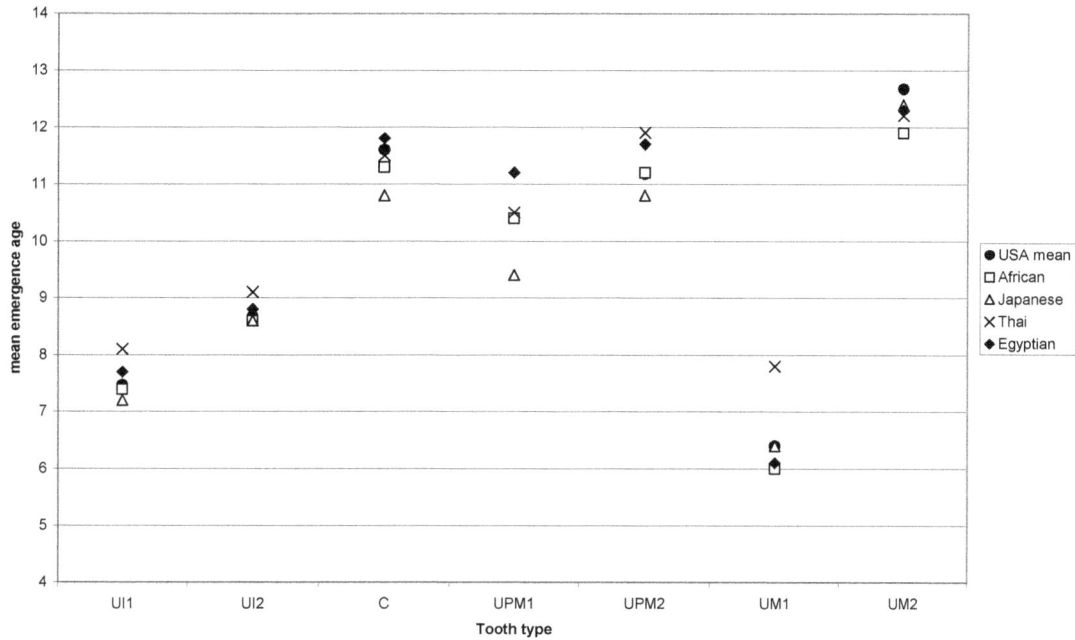

Figure 1. Comparison of mean age of tooth emergence in world populations

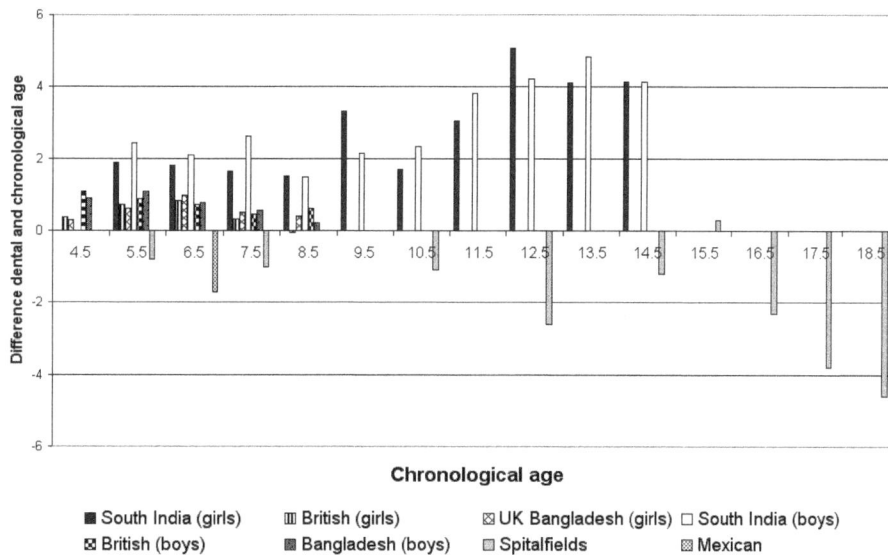

Figure2. The difference between age predicted using Demirjian et al.'s system and known chronological age.

estimates are compared to the developmental stage of the Spitalfield children the archaeological sample is almost exclusively younger than chronological age. This pattern is only matched by the Mexican children. All other populations are estimated as older than their chronological age. One factor here is the comparison between mean data and individuals. However, the data from the French Canadian reference sample shows how great the range can be in a population.

Skeletal development
The skeletal indicators appear to be a more accurate

predictor of age than dental indicators, contrary to the usual expectations (Figure 3). All the children have less than a two years difference between their chronological age and the estimated skeletal age. Skeletal age however, underestimates age for almost all the children, so based on skeletal aging they would look slightly younger than their known age. Although some of the largest differences exist in the older children, skeletal age is more accurate for these children than dental age, with less than half the difference exhibited by dental age. This should not be surprising, as after the 2nd molar has developed the teeth can no longer be relied on for age prediction (Eveleth & Tanner 1990).

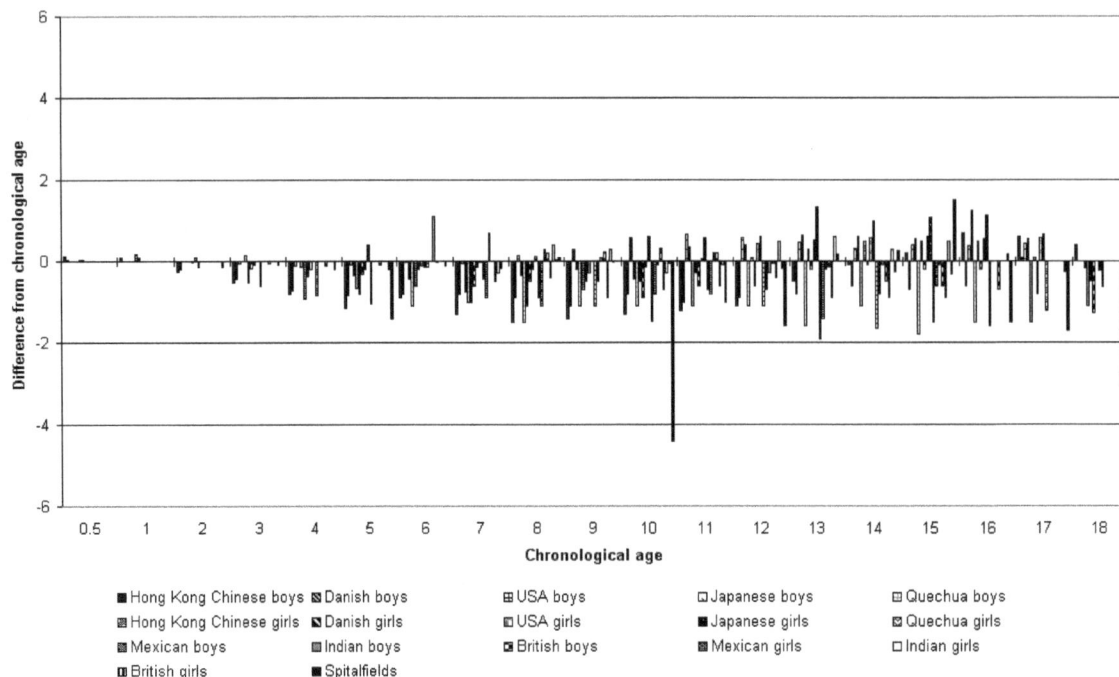

Figure 3. The difference between skeletal age and known chronological age using TW2.

Discussion

The range of variation in skeletal and dental maturation within a population is usually quite wide. Children of the same chronological age can have widely differing physiological ages. Conversely, children of the same physiological age may have been born years apart. This has been noted in the growth and development literature by many authors, for example, Lampl & Johnston (1996) in Mexican children and by Eveleth & Tanner (1990) in their book on this subject.

The data for tooth emergence used on the mean age from the various populations. It would have been useful to compare the range of variation within and between these populations. However, many of the studies used in this paper are quite old, the data was published long ago and sometimes in obscure journals. This makes it difficult to obtain, the WHO is currently compiling new standards for different ethnic groups which should enable a greater insight into the similarities and differences between groups. The differences and similarities indicated by the mean data may, as Eveleth and Tanner suggest, be more socio-economic than ethnic. Ill health and malnutrition may exaggerate minor variations.

The use of reference populations for either dental or skeletal age estimates has produced differences between known chronological age and physiological age in all populations (Figures 1 & 2). In other words no other population as the same mean developmental rate for either tooth emergence, tooth development or bones as the reference populations. However, if the range of variation is used even when drawn from this same single reference population, then most other populations fall within this range. This study has shown the wide variation in mean maturation rates in different regions, when compared to published standards. A single child therefore, is unlikely to have an exact match between chronological and physiological age. This would be true even where the child from the reference population, in which the age difference between the two can be up to eighteen months (Demirjian *et al.* 1976; Jones *et al.*, 1973).

This mismatch is illustrated by the inclusion of the Spitalfields children. These children are mostly older than their maturation stage would indicate (Clegg & Aiello, 1999). The difference could be argued to relate to poorer healthcare in the past or because these children died young (Molleson & Cox, 1993; Liversidge & Speechley, 2001). This is discussed more fully by Clegg & Aiello, (1999) and Molleson and Cox, (1993). There may also be something different about the growth in children who die young compared to those who live to grow up. Unfortunately, the normal children from these past populations are those we find as adults in our archaeological samples. We could therefore argue that they might not represent their populations in respect of past growth and development. However, they are all we have and we must find an adequate solution to our problem.

Most studies rely on dental development to assess age at death; however, this study has shown that contrary to expectations, skeletal development might be closer to chronological age, particularly for older individuals. However, the same provisos still hold; there is no exact match with chronological age.

We could overcome this by comparing archaeological populations with the maturation rates of modern children who suffer long term ill-health or who die young. This might overcome some of the problems, as children with know diseases do differ from their more normal counterparts (Kong *et al.* 1999; De Luca & Baron 1999; Henderson *et al.* 2005). However, different populations vary in the timing of developmental stages and so the differences may be influenced more by socio-economic, climatic or other ecological factors, we might therefore try to match children on these criteria (Lampl & Johnson 1996; Tompkins 1996). This will soon be possible as countries round the world develop their own growth and development standards. We might also want to develop growth standards for archaeological samples. To do this we would need to undertake a large scale study comparing known age skeletal samples with known age living children. Both these aims might be our long term goal. In the short term however, we should acknowledge that few children will be at the mean age for development stage either dentally or skeletally. We must stop striving for greater and greater precision, and acknowledge that past populations and the children within them would show the range of variation in their maturation rates exhibited by present day populations. To say that a child is within a certain age range, based on their teeth or bones acknowledges this variation and prevents us from proposing, contrary to the evidence, that we can or should do more.

References

Clegg M & Aiello LC. 1999. A comparison of the Nariokotome *Homo erectus* with juveniles from a modern human population *American Journal of Physical Anthropology* 110 (1) 81-94

De Luca F & Baron J. 1999. Skeletal maturation. *Endocrinologist.* 9:4:286- 293

Demirjian A & Goldstein H. 1976. New systems for dental maturity based on four and seven teeth *Annals of Human Biology* 3, 411-421

Demirjian A. 1986. Dentition. In Falkner, F. & Tanner, J.M. (Eds.) *Human Growth: A comprehensive treatise* 2nd ed. Vol. 2 New York Plenum Press pp260-298

Eveleth PB & Tanner JM. 1976. *World-wide variations in human growth.* 1st ed. Cambridge: Cambridge University Press

Eveleth PB & Tanner JM. 1990. *World-wide variations in human growth.* 2nd ed. Cambridge: Cambridge University Press

Garn SM, Rohmann CG, Silverman FN. 1967. Radiographic standards for postnatal ossification and tooth calcification. *Medical Radiophotography.* 43 45-65

Henderson RC, Gilbert SR, Clement ME, Abbas A, Worley G, Stevenson RD. 2005. Altered skeletal maturation in moderate to severe cerebral palsy. *Developmental Medicine and Child Neurology.* 47:4:229-236

Hoppa RD & Fitzgerald CM. 1999. *Human Growth in the past.* Cambridge: Cambridge University Press.

Kong CK, Tse PWT, Lee WY. 1999 Bone age and linear skeletal growth of children with cerebral palsy. *Developmental Medicine and Child Neurology* 41:11:758-765.

Krogman WM & Iscan MY. 1996. *The human skeleton in forensic medicine.* Springfield IL: C. Thomas.

Lampl M & Johnston FE. 1996. Problems in the ageing of skeletal juveniles: Perspectives from maturation assessments of living children. *American Journal of Physical Anthropology* 101:345-355

Liversidge HM & Speechly T. 2001. Growth of permanent mandibular teeth of British children aged 4 to 9 years. *Annals of Human Biology.* 28:3:256- 262

Molleson T & Cox M. 1993. *The Spitalfields. Project Vol. 2 The Anthroplogy: The Middling Sort.* CBA Research Report 86 London: Council for British Archaeology.

Nykanen R, Espeland L, Kvaal SI, Krogstad O. 1998. Validity of the Demirjian method for dental age estimation when applied to Norwegian children. *Acta Odontol Scand.* 56:238-244.

Satake T. 1999. Sexual Dimorphism in the Relationship Between Number of Emerged Permanent Teeth and Percentage of Adult Stature. *American Journal of Human Biology* 11:619–626

Smith BH. 1991. Standards of human tooth formation and dental age assessment. In Kelley MA and Larsen CS. (eds.) *Dental Anthropology.* New York. Wiley-Liss Inc.

Tompkins RL. 1996. Human Population Variability in Relative Dental Development. *American Journal of Physical Anthropology* 99:79-102.

Tuberculosis at Spitalfields, London:
an initial insight into medieval urban living.

A. Gray Jones & D. Walker

Abstract

A group of 10,500 skeletons were archaeologically excavated from the cemetery of the medieval hospital and priory of St Mary-without-Bishopsgate (Spitalfields), London, England. The osteological analysis of the sample is ongoing until 2006 and this paper aims to report some of our initial findings, with particular reference to the prevalence of tuberculosis in the medieval population. In the sample of 1,654 skeletons studied so far, twenty-eight individuals were found to have the skeletal signs of tuberculosis (a crude prevalence rate of 1.7%). Initial trends suggest that the greatest prevalence was found in teenage and young adult individuals (12-25 years of age) and that more males were affected than females. The study of this un-paralleled sample from an urban medieval population has the potential to allow for a better understanding of tuberculosis throughout the entire life-course. In the future this bioarchaeological data, considered in combination with an un-rivalled understanding of the living conditions in medieval London, as derived from a wealth of literary and archaeological sources, will provide a unique opportunity to examine the prevalence and effects of this infectious disease within the urban environment.

Keywords: Spitalfields; Tuberculosis; London; Medieval

Introduction

Tuberculosis is a chronic infectious disease caused by *Mycobacterium tuberculosis* and *Mycobacterium bovis*. *M. tuberculosis* is transmitted via droplet infection from human to human, whilst *M. bovis* is transmitted via the ingestion of meat and milk from animals, especially cattle, or via droplet infection. It is primarily a disease of the soft tissues with only a minority of individuals showing skeletal involvement and identifiable in palaeopathology; data from the preantibiotic era suggest bone changes are seen in only about 5-7% of cases (Steinbock 1976).

Although there is abundant palaeopathological evidence for the disease from around the world, the excavation of over 10,500 skeletons from a medieval hospital in Spitalfields, London, represents a unique opportunity for the study of tuberculosis within an urban population. Not only does such a large and well-dated sample provide a wealth of bioarchaeological data, it is supported by an un-rivalled understanding of the living conditions in medieval London, derived from literary and archaeological sources.

Spitalfields and St. Mary Spital are the common names for the medieval priory and hospital of St. Mary-without-Bishopsgate, which was founded in response to the need for new hospitals to care for pilgrims, the poor and the sick of London's growing population. In use between 1197 and 1539, the priory's cemetery spans an important time in the development of London, a time in which the city was experiencing extraordinary growth. Recent research has suggested that the population doubled during both the 12th and 13th centuries, with a population of around 80-100,000 by 1300 (Keene 1984), perhaps more than twice that of other cities in England (Thomas 2002).

Archaeological recording by the Museum of London Archaeological Service (MoLAS) took place between 1999 and 2002, ahead of redevelopment of the area, and was one of the largest urban excavations ever undertaken, covering 5.2 hectares. The archaeology on the site continues from the Roman through to the post-medieval period and for the medieval period alone the excavations included the priory buildings, charnel house and approximately 10,500 skeletons from the cemetery.

There was remarkable homogeneity in burial practice - the vast majority of individuals were laid out in single graves, aligned east-west, with the head at the west end and the body laid supine. Approximately 2000 individuals however were found in mass burial pits containing multiple burials up to five layers deep with as many as 45 individuals in a single pit. Radiocarbon dating has shown that these people were buried at some time between 1270 and 1320, about half a century earlier than the first visit of the Black Death to England in 1348-9. Dating of the archaeological phases of the site is ongoing and a preliminary study using radiocarbon dates to test the phasing has shown it to be very reliable.

Post-excavation work on the osteological material has included an assessment of all the human bone and a pilot project in which 200 individuals were fully studied. Around 6,800 individuals were identified for full osteological analysis, based on a minimum requirement of 35% completeness for each skeleton. This paper is based on the analysis of a total of 1,654 skeletons studied by the time the presentation was prepared. As the analysis of both the skeletal material and the spatial and temporal phasing of the archaeology is ongoing the sample discussed here is essentially random, spanning all of the medieval phases of the cemetery and regardless of burial type: i.e. attritional or mass burial. In this paper we aim to present our preliminary results in order to give an initial insight into the prevalence of tuberculosis in the population of medieval London.

Osteological methods

A sample of 1,654 of the skeletons selected for study has been examined osteologically, 28 of these show signs of skeletal tuberculosis and are the subject of this paper. Tuberculosis was diagnosed macroscopically according to the morphology of the lesions and their distribution throughout the skeleton, using the diagnostic criteria presented by Aufderheide & Rodriguez-Martin (1998), Ortner (2003), and Resnick (2002). The main characteristics included: predominantly destructive lesions affecting the vertebral bodies or long-bone metaphyses; a predilection for the lumbar and thoracic vertebral bodies and the large joints of the appendicular skeleton; little reactive new bone formation; and frequent vertebral collapse with kyphosis of the spine. Rib lesions, consisting of active/remodelling new bone and/or erosive lesions, were considered to be an extension of the spinal foci where they were confined to the rib head/neck and/or adjacent to affected vertebrae. In cases where rib lesions were distributed over all or part of the whole rib shaft, and not associated with affected vertebra, they were considered possibly to represent sub-adjacent pulmonary infection.

The sexing of adult skeletons was based on visual observation of morphological characteristics of the skull and pelvis and scored on a five point scale (Buikstra & Ubelaker 1994). Sex determination was not attempted for individuals under 18 years of age.

Age estimation of adults was based on morphological changes on the pubic symphysis (Brooks & Suchey 1990), the auricular surface (Lovejoy et al 1985) and at the costo-chondral junction (Iscan et al 1984:1985). In addition the amount of wear on the mandibular molar teeth (Brothwell 1981) was employed as a broad guide to the age of death. For sub-adults, age determination was based on lengths of long bone diaphyses (Scheuer et al 1980), on stages of dental development (Moorrees et al 1963a, 1963b) and tooth eruption (Gustafson & Koch 1974), and on the overall progress of epiphyseal fusion (Scheuer & Black 2000).

Results

Twenty-eight individuals were found to have the skeletal signs of tuberculosis (Table 1).

The distribution of skeletal lesions consisted largely of active lesions in the spine. These were focused on the centra of the thoracic and/or lumbar vertebrae and often resulted in collapse and kyphosis. In three cases the sacral vertebrae were affected in addition to thoracic or lumbar lesions and in one individual the cervical vertebrae were also affected. In approximately one third of cases, the posterior elements of the vertebrae were affected in addition to lesions of the centrum. Eleven individuals (approximately 40% of those with tuberculosis) also had new bone formation on the visceral surfaces of the ribs, and in five cases were accompanied by erosive lesions. In all of the individuals the location and morphology of these rib lesions suggested that they represented extension from the spinal foci. However four of these individuals were thought to have a combination of lesions; both those spread from the spinal foci and new bone formation/erosive lesions which may represent sub-adjacent pulmonary infection. Some of the pathological changes are illustrated in Figure 1.

TABLE 1. TUBERCULOSIS PREVALENCE RATES SHOWN AS A PERCENTAGE OF THOSE IN EACH SEX AND AGE CATEGORY.

Age category (years)	Sub-adult % (n/total)	Male % (n/total)	Female % (n/total)	Total % (n/total)
6-11	1.6 (2/123)			1.6 (2/123)
12-17	2.3 (4/172)			2.3 (4/172)
18-25		3.2 (5/157)	0.9 (1/109)	2.3 (6/266)
26-35		1.7 (4/234)	2.6 (5/194)	2.1 (9/428)
36-45		1.9 (4/212)	1.5 (2/131)	1.7 (6/343)
> 46		1.4 (1/70)	0 (0/53)	0.8 (1/123)
Total	2.0 (6/295)	2.1 (14/673)	1.6 (8/487)	1.9 (28/1455)

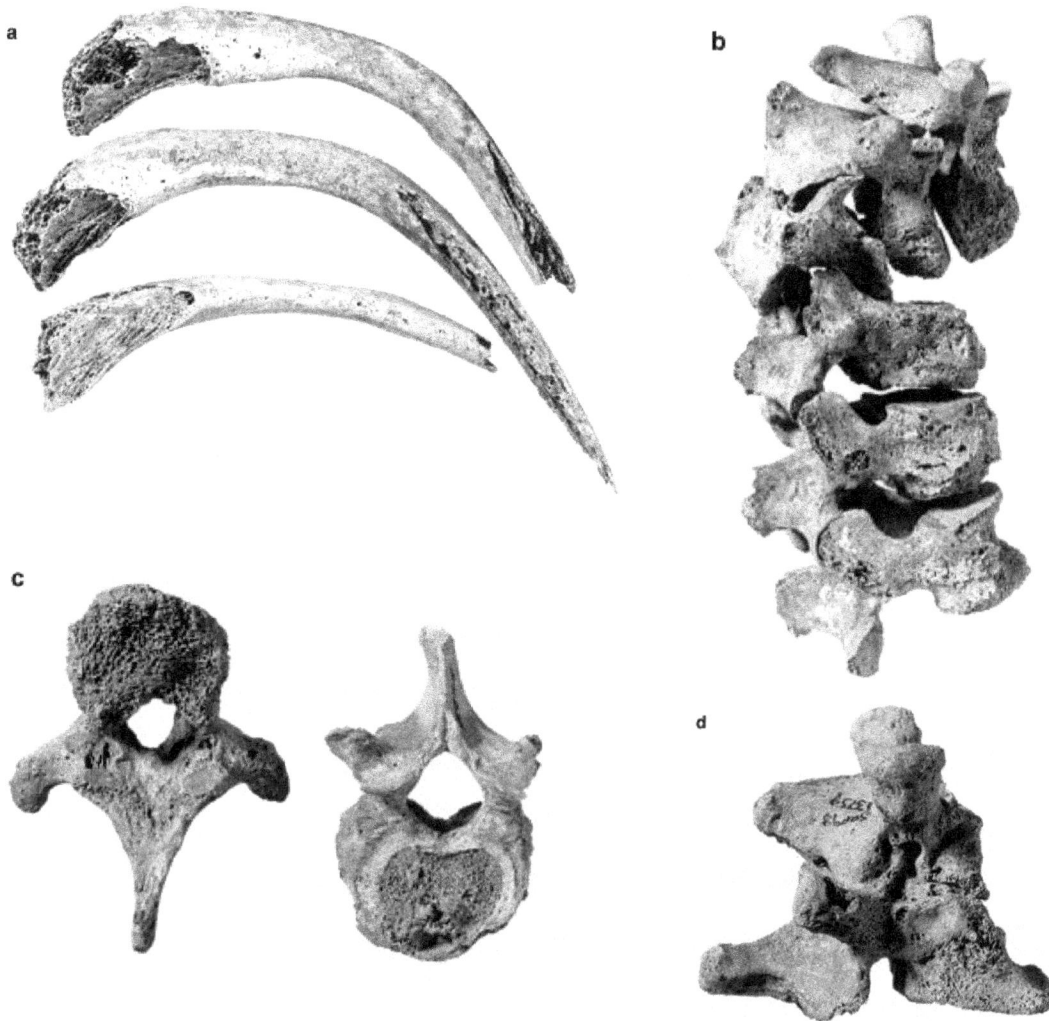

Figure 1. Examples of osseous lesions suggestive of tuberculosis in some of the Spitalfields skeletons.

a: Skeleton 22829. Proliferative and erosive visceral rib lesions. b: Skeleton 22829. T9-L2. Collapse of the 9th to 11th thoracic vertebrae due to destruction of the centra and erosive lesions on the inferior and superior articular facets of T9-T11. c: Skeleton 29490. 7th and 10th thoracic vertebrae showing destructive lesions on the inferior and superior centrum. d: Skeleton 13759. 10th and 11th thoracic vertebrae showing complete destruction of the centra.

There were two individuals with involvement of the sacroiliac joint, consisting of localised destructive lesions with no blastic response, but no other joints were affected. In one individual, in addition to lesions on the vertebrae and ribs, the lesser trochanter of the left femur had been largely destroyed by a lytic lesion. The greater trochanter is known as a focus for tuberculous lesions, originating in the bursa or the bone (Ortner 2003:239), but in this case it was damaged post-mortem. It is possible that the lesser trochanter may be affected in a similar way and as the attachment for the psoas muscle, it was also considered that the infection may have spread from the spinal foci to the femur via the psoas muscle.

The crude prevalence rate found within the sample so far is 1.7% of individuals (28/1654), or 1.7% of all adults (22/1277) and 1.6% of sub-adults (6/377). The age and sex distribution of the sample studied is shown in Figure 2, and the prevalence of tuberculosis within the sample is shown in Figure 3. There were several individuals in the current sample where the determination of sex could not be observed, or was deemed intermediate, and where age at death could only be determined as adult or sub-adult. This data has not been included in our current analysis, other than for the crude prevalence rates. None of these individuals showed skeletal signs of tuberculosis and there were no cases of tuberculosis in sub-adults younger than six years of age. Consequently sub-adults younger than six years (n=81) are also excluded from Table 1 and therefore the total considered is 1,455.

Demographic distribution of the sample

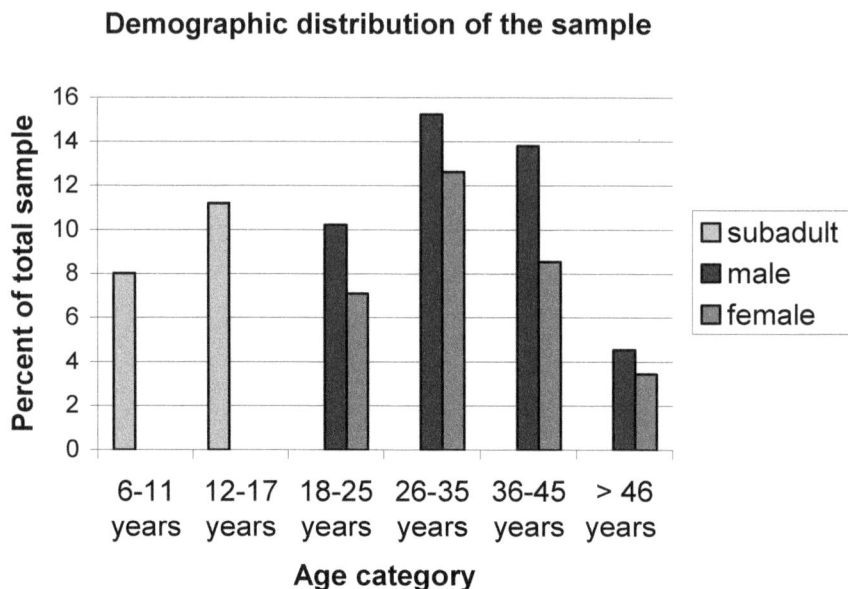

Figure 2

Prevalence of tuberculosis in the sample

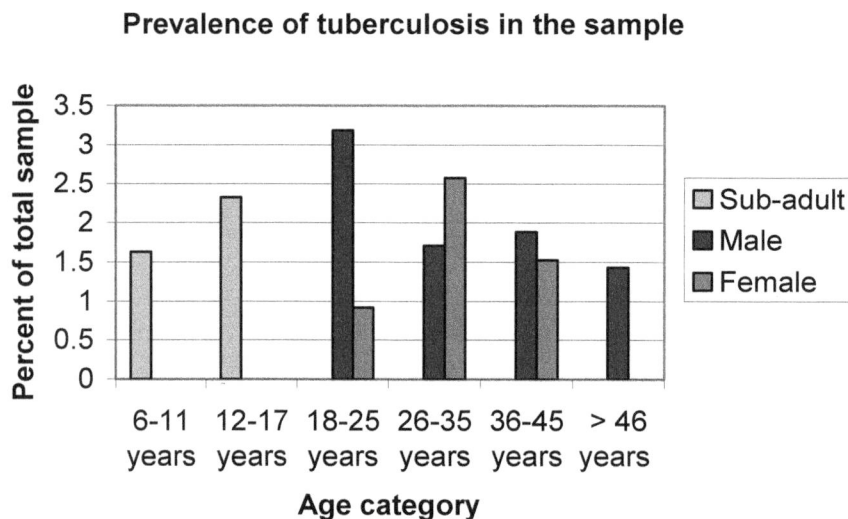

Figure 3

Examination of these initial results reveals some general trends in the data. There appears to be a peak in the prevalence of tuberculosis between 12 and 25 years of age (affecting 2.3% of both 12-17 year olds and 18-25 year olds). The rate appears to increase amongst sub-adults to reach this peak, and then to decrease, at first gradually and then more rapidly, with increasing age. There were no cases of tuberculosis in sub-adults younger than six years of age.

There also appears to be a difference in the prevalence of tuberculosis between the sexes, especially in the 18-25 year age group, when a greater number of males affected. This difference appears to lessen with increasing age, but not before a reversal of the trend, with a slightly higher

prevalence of tuberculosis in females than in males in the 26-35 year age category. Although 1.4% (1/70) of the older males (>46 years) were affected none of the 53 older females recorded in the sample showed signs of tuberculosis.

The apparent differences in the prevalence of tuberculosis between males and females, and between age categories, were subjected to statistical validation using the chi-square test (with Yates' correction used when there were only two categories of data, i.e. one degree of freedom). The patterns were not found to be statistically significant.

Discussion
The principle purpose of this paper is to provide an initial

indication of the general pattern of prevalence of tuberculosis within the population at this early stage of analysis. The osteological recording of the cemetery sample is ongoing until 2006 and should be considered as a work in progress.

Although we have been unable to demonstrate it at a significant level here it would be wrong to assume that tuberculosis affected all age groups uniformly. There are a number of factors which are known to influence the prevalence of disease within a population. The nature of palaeopathology, for example, may explain a paucity of evidence in older individuals, younger individuals survive to the chronic stages of infection (therefore showing the bony changes) while older individuals are less able to resist the acute stages and therefore would be under-represented in the sample. This may also be true for younger sub-adults, with a lack of cases observed in children younger than six years of age. Alternatively, clinical evidence from modern populations has shown that in the early stages of an epidemic the disease tends to be found in the younger members of the population, with most people affected by the age of 20 (Resnick 2002:2525). A higher prevalence of tuberculosis in younger age categories may therefore reflect an epidemic in its early stages. However it is also true that skeletal tuberculosis has a higher incidence amongst juveniles than adults when the pulmonary form of the infection prevails in a population (Jane Buikstra *pers comm*).

Beyond this, there may have been differences in the prevalence of tuberculosis between the sexes, as has been shown in modern populations (WHO Report 2002). To some extent this may reflect a difference in immune status, with females having a stronger immune system and thus being more able to live longer with the disease (Ortner 1998). Specifically, it has been suggested that female health may be more resilient during late adolescence/early adulthood than male health (Stinson 1985).

These biological aspects must also be considered in the light of socio-cultural and environmental factors that may influence the prevalence of tuberculosis. An individual's sex and/or age may determine their occupation, geographical mobility and the degree and range of their social interaction, all of which would vary the opportunity to contract and spread the disease. In skeletal samples we can observe variations in the prevalence of disease between the sexes and as well as biological reasons for these we must also consider that the observed differences could reflect gender differences which can also influence the risk of contracting or spreading the disease.

In addition, varying attitudes to the disease and to illness in general would have an effect on diagnosis and when, or whether, treatment took place. It is also probable that different social groups would have had varying levels of access to healthcare and hygiene, much as has been demonstrated that they do now (WHO Report 2002). Differences in the local environment, such as living in close proximity to animals and to other people, would affect general living conditions, as well as the risk of contracting tuberculosis.

All the above factors will often vary with both age and gender, as different groups are subject to different social or cultural traditions. These influences are dynamic and will change throughout the entire course of an individual's life. One possible scenario which may demonstrate how these factors could combine is the migration of children, possibly as young as seven years, and young adults, particularly males, from rural to urban environments seeking work (Pelling 1994). Migration to a new environment is thought to increase the risks of both the exposure to, and the contraction of, tuberculosis (Roberts & Buikstra 2003). Immigrants often suffer from poor living conditions, with overcrowding, low standards of hygiene and poor sewerage disposal and this environmental stress, combined with a depressed immune system when compared to local inhabitants, would increase the risk of tuberculosis (Roberts & Buikstra 2003). Migration may also expose individuals to new pathogens or increase the risk of re-infection or re-activation of a dormant infection, especially as migrants often live and/or work in close proximity to each other and socialise within their own group. Overall, migration makes these individuals more susceptible to infection, and makes transmission within their social group more likely.

Comparisons with other sites
We can compare the initial, crude, prevalence rates from Spitalfields with contemporary samples from both rural and urban contexts. The Royal Mint site in the City of London had a lower prevalence rate of 0.5% (Roberts & Cox 2003), which may reflect differences in the population buried there, such as a generally higher standard of living with less cramped conditions, better sanitation and diet, and access to healthcare.

In York, the Jewbury cemetery and St. Andrew Fishergate both had a prevalence rate of approximately 1.5% (Roberts & Cox 2003). These rates are only marginally lower than at Spitalfields, and imply that conditions in towns during this period enabled the disease to be effectively transmitted within the urban population. A third cemetery in York, St. Helen-on-the-Walls, had a prevalence rate of only 0.5% (Roberts & Cox 2003), similar to that at the Royal Mint in London, which again may reflect varying living conditions within the city's population. Given that St. Helen-on-the-Walls has been identified as one of the poorest areas of York, it is possible that the low prevalence of skeletal tuberculosis reflects a population that was poorly equipped to survive the acute stages of the disease.

The prevalence of tuberculosis in the contemporary rural sample from Wharram Percy in Yorkshire was 1.3% (Mays 2002), lower than that at Spitalfields. However, the prevalence amongst adults was actually higher. Biomolecular analysis has shown that this was *Mycobacterium tuberculosis,* not the form of tuberculosis contracted from cattle (*M. bovis*).

Implications for the study of medieval urban life
It is well established that tuberculosis is a disease of poverty and urban environments, characterised by unhygienic and overcrowded conditions. Although no statistically significant differences have been found at this stage our initial results suggest that whilst this may be the case, the prevalence of tuberculosis may also vary between age and sex groups. Reasons for differences such as these include the introduction of specific groups, such as young male migrant workers, but the situation is likely to be more complex. The results from Wharram Percy, for example, show that tuberculosis was not just an urban disease and was not simply transmitted via infected cattle in rural contexts and from person to person in urban contexts (Mays 2001).

Both age and gender divisions affect occupation, mobility and social interaction throughout an individuals life, varying the likelihood of contracting, transmitting and surviving the disease. Modern studies by public health organisations have found differences between the sexes in rates of infection, diagnosis and access to treatment (WHO Report 2002). The challenge within palaeopathology, and in the analysis of the Spitalfields cemetery population, is to relate these cultural factors to the archaeological prevalence rates.

Further work
The Spitalfields cemetery provides a unique opportunity to study a large and well dated sample of a medieval city's population. The final analysis of tuberculosis will be based on real prevalence rates (the number affected as a percentage of those with relevant surviving skeletal elements) and the type and distribution of lesions will be examined in more detail. This will include an investigation of the link between rib lesions and pulmonary tuberculosis, and the relationship between tuberculosis and other diseases and indicators of biological stress.

The phasing of the cemetery will allow us to study the prevalence of tuberculosis over time, and to compare the health of those buried in the attritional cemetery with those in the mass burial pits.

Finally, ongoing lead isotope analysis (in collaboration with Reading University) may identify the geographic origins of a sample of the skeletons, indicating any recent migrants within the population. Further potential exists for biomolecular analysis of those individuals with skeletal tuberculosis, to both confirm the palaeopathological diagnosis and to indicate whether infection was by *M. tuberculosis* or *M. bovis*.

To conclude, being chiefly a disease of soft tissue, the prevalence rates of tuberculosis in this study represent only a minimum indication of the impact of the disease on the population. Tuberculosis must have seriously compromised public health, and we know that by the 17th century it was responsible for approximately 20% of all deaths in London (Clarkson 1975). Given that tuberculosis has been designated a "global emergency" by the World Health Organisation it is clear that we can not overestimate the serious risk that it held in past societies. It is hoped that as our analysis progresses over the next two years we will develop a further understanding of the nature and prevalence of tuberculosis in medieval London.

Acknowledgements
Firstly our thanks go to Rebecca Redfern, Natasha Powers and Brian Connell, our co-workers on the Spitalfields Osteology project, whom whilst having contributed to the data upon which this paper is based have also provided invaluable support and advice during the preparation of this manuscript. Thanks are also due to Jane Buikstra for her supportive comments on the presentation manuscript and to Chris Thomas, and the Museum of London Archaeology Service, for assistance in attending the conference. The Spitalfields Project has been funded principally by the Spitalfields Development Group (a subsidiary of Hammerson Plc.). All photos are by Andy Chopping and Maggie Cox, Museum of London Archaeology Service.

References

Aufderheide AC & Rodriguez-Martin A. 1998. *Encyclopedia of Human Paleopathology*. Cambridge University Press: Cambridge.

Brooks ST & Suchey JM. 1990. Skeletal Age Determination Based on the Os Pubis: comparison of the Ascadi-Nemeskeri and Suchey-Brooks methods. *Human Evolution* 5:227-238

Brothwell DR. 1981. *Digging Up Bones*. 3rd Edition BM(NH) and OUP: London and Oxford.

Buikstra JE & Ubelaker DH (Eds). 1994. *Standards for Data Collection from Human Skeletal Remains*. Arkansas Archaeological Survey Research Series, no. 44: Fayetteville

Clarkson L. 1975. *Death disease and famine in pre-industrial England*. Dublin.

Gustafson G & Koch G. 1974. Age estimation up to 16 years of age based on dental development. *Odontologisk Revy* 25: 297-306

Iscan MY, Loth SR, Wright RK. 1984. Age estimation from the rib by phase analysis: white males. *Journal of Forensic Sciences* 29:1094-1104.

Iscan MY, Loth SR, Wright RK. 1985. Age estimation from the rib by phase analysis: white females. *Journal of Forensic Sciences* 30:853-863

Keene D. 1984. A new study of London before the Great Fire. *Urban History Yearbook* 1984.

Lovejoy CO, Meindl RS, Pryzbeck TR, Mensforth RP. 1985. Chronological metamorphosis of the auricular surface of the ilium: a new method for the determination of adult skeletal age at death. *American Journal of Physical Anthropology* 68:15-28

Mays S, Taylor GM, Legge AJ, Young DB, Turner-Walker G. 2001. Palaeopathological and biomolecular study of tuberculosis in a medieval skeletal collection from England. *American Journal of Physical Anthropology* 114:298-311.

Mays S, Fysh E, Taylor GM. 2002. Investigation of the link between visceral surface rib lesions and tuberculosis in a medieval skeletal series from England using ancient DNA. *American Journal of Physical Anthropology* 119:27-36.

Moorrees CFA, Fanning EA, Hunt EE Jr. 1963a. Formation and resorption of three deciduous teeth in children. *American Journal of Physical Anthropology* 21:205-213

Moorrees CFA, Fanning EA, Hunt EE Jr. 1963b Age variation of formation stages for ten permanent teeth. *Journal of Dental Research* 42(6):1490-1502

Ortner DJ. 1998. Male-female immune reactivity and its implications for interpreting evidence in human skeletal palaeopathology. In *Sex and gender in palaeopathological perspective*, Grauer AL & Stuart-Macadam P. (eds.). Cambridge University Press: Cambridge.

Ortner DJ. 2003. *Identification of Pathological Disorders in Human Skeletal Material*. Academic Press: London.

Pelling M. 1994. Apprenticeship, health and social cohesion in early modern London. Hist Workshop J 37:33-56.

Resnick D. 2002. (ed.). *Diagnosis of Bone and Joint Disorders*. WB Saunders: Philadelphia.

Roberts CA & Buikstra JE. 2003. *The Bioarchaeology of Tuberculosis. A Global View on a Re-emerging Disease*. University of Florida: Florida.

Roberts CA & Cox M. 2003. *Health and Disease in Britain. From Prehistory to the Present Day*. Sutton Publishing: Stroud.

Scheuer JL, Musgrave JH, Evans SP. 1980. The estimation of late foetal and perinatal age from limb bone length by linear and logarithmic regression. *Annals of Human Biology* 7(3):257-265

Scheuer L and Black S. 2000. *Developmental Juvenile Osteology*. Academic Press: London.

Steinbock RT. 1976. *Palaeopathological diagnosis and interpretation*. C.C. Thomas: Springfield, IL.

Stinson S. 1985. Sex differences in environmental sensitivity during growth and development. *Yearbook of Physical Anthropology* 28:123-147.

Thomas C. 2002. *The archaeology of medieval London*. Sutton Publishing: Stroud.

World Health Organisation. *Global Tuberculosis Control*. WHO Report 2002. Geneva, Switzerland, WHO/CDS/TB/2002.

KLIPPEL-FEIL SYNDROME: EXAMPLES FROM TWO SKELETAL COLLECTIONS OF ALASKAN NATIVES

S.S. Legge

Abstract

Vertebral fusion was observed as part of a larger study of spinal pathologies in skeletal collections of Alaskan Natives from Golovin Bay and Nunivak Island, USA. Block vertebrae that were clearly congenital were differentiated from vertebral fusion resulting from osteophyte development or other pathological conditions. Congenital fusion (segmentation failure) was present in the cervical vertebrae of two individuals from each collection. The pathologies observed most likely represent cases of Klippel-Feil syndrome. Three cases exhibit segmentation failure of the second and third cervical vertebrae, while the third and fourth cervical vertebrae are connected in the fourth individual. The frequencies of segmentation failure between the second and third cervical vertebrae are 1.9% for Golovin Bay and 4.7% for Nunivak Island. These are some of the first cases of this condition observed in Alaskan Natives and will hopefully add to our understanding of the occurrence of this rare congenital malformation among past populations in Alaska.

Keywords: Klippel-Feil syndrome; vertebral pathology; Alaskan Natives

Introduction

Vertebral fusion falls into two main categories, congenital and pathological. The most common type of vertebral fusion is the result of pathological conditions that stimulate osteophyte development (Schmorl & Junghanns 1971). However, fusion of the vertebral column in skeletal collections has also been attributed to developmental defects, such as Klippel-Feil syndrome (Ortner & Putschar 1985). Klippel-Feil syndrome involves two or more cervical vertebrae connected as a solid block (Aufderheide & Rodríguez-Martín 1998). The aetiology of this condition is believed to be genetic, although the exact method of heredity has not been determined (Aufderheide & Rodríguez-Martín 1998). Block vertebrae that are congenital in origin are not actually fused, but rather they failed to separate in the developing vertebral column (Barnes, 1994). Cases of segmentation failure of cervical vertebrae have been discussed in a few skeletal collections (Wade 1981; Merbs & Euler 1985; Danforth *et al.* 1994; González-Reimers *et al.* 2001; Merbs 2004), although the overall frequency of occurrence in archaeological context is not well documented. There are a variety of reasons why this may be the case. It is possible that pathologies of the lower vertebral column have a higher survivability rate than those of the upper vertebral column. Additionally, differential archaeological preservation may even play a role in the recovery, and therefore observation, of various pathological conditions. This paper examines four

possible cases of Klippel-Feil syndrome in two skeletal collections of Alaskan Natives. It is hoped that observations of these defects will provide some insight into the frequency of occurrence of this condition among earlier peoples of Alaska.

Materials and Methods

Vertebral fusion was observed as part of a larger study of spinal pathologies in two Eskimo skeletal collections from different geographic regions in Alaska. The Golovin Bay skeletal collection came from the south side of the Seward Peninsula (Figure 1).

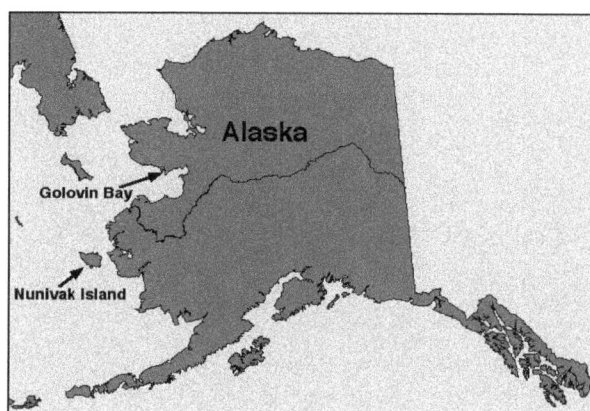

Figure 1. Locations of Nunivak Island and Golovin Bay.

The second collection was from Nunivak Island, located approximately 40 miles off of the southwest coast of Alaska (Figure 1). The samples were collected and accessioned by the Smithsonian Institution between 1907 and 1931. The archaeological ages of the skeletal collections were not available, however, it was noted that the majority of the remains from Nunivak Island most likely dated to the late 19th and early 20th centuries (Speaker *et al.* 1996). All vertebrae were observed and documented utilizing the Smithsonian protocol for skeletal analysis (Verano & Urcid 1994). This protocol was developed by the Repatriation Office of the National Museum of Natural History at the Smithsonian Institution. The age and sex of each individual was estimated using criteria set forth in the protocol. All analysis was performed at the University of Alaska Fairbanks prior to repatriation.

All block vertebrae found above the lumbosacral junction are noted and described according to which vertebrae are involved and the portions of the vertebrae fused. Where possible, an attempt is made to ascertain the cause of the fusion in each case, such as congenital versus the result of osteophyte development, in order to better describe the nature of this condition in each collection. Fusion was classified as congenital if the following conditions were met: extensive osteophyte development was not observed; there was no clearly related healed trauma; and any osteoarthritic remodelling appeared to be secondary to the fusion.

There are at least 165 individuals in the Golovin Bay collection and 139 individuals in the Nunivak collection. Following Merbs (2004) frequencies were calculated based on the occurrence of segmentation failure for each C2/C3 and C3/C4 intervertebral space. At Golovin, 54 and 46 individuals were observable for C2/C3 and C3/C4 respectively (Table 1). For the Nunivak collection the numbers of individuals available were 43 for C2/C3 and 26 for C3/C4 (Table 2). The most common bone in both collections is the cranium. Most individuals are represented by only a few vertebrae and many have none at all.

TABLE 1. NUMBERS OF INDIVIDUALS WITH C2/C3 AND C3/C4 INTERVERTEBRAL SPACES REPRESENTED IN THE GOLOVIN BAY COLLECTION.

Age (years)	Male		Female		Indeterminate		Totals	
	C2/C3	C3/C4	C2/C3	C3/C4	C2/C3	C3/C4	C2/C3	C3/C4
Newborn – 0.9	-	-	-	-	1	0	1	0
1 – 4	-	-	-	-	1	1	1	1
5 – 9	-	-	-	-	2	1	2	1
10 – 14	-	-	-	-	3	3	3	3
15 – 19	1	1	0	0	3	1	4	2
20 – 34	6	4	11	10	0	0	17	14
35 – 49	9	8	12	12	0	0	21	20
50 +	1	1	3	3	0	0	4	4
Unknown adult	0	0	1	1	0	0	1	1
Totals	17	14	27	26	10	6	54	46

TABLE 2. NUMBERS OF INDIVIDUALS WITH C2/C3 AND C3/C4 INTERVERTEBRAL SPACES REPRESENTED IN THE NUNIVAK ISLAND COLLECTION.

Age (years)	Male		Female		Indeterminate		Totals	
	C2/C3	C3/C4	C2/C3	C3/C4	C2/C3	C3/C4	C2/C3	C3/C4
Newborn – 0.9	-	-	-	-	0	0	0	0
1 – 4	-	-	-	-	1	0	1	0
5 – 9	-	-	-	-	0	0	0	0
10 – 14	-	-	-	-	2	1	2	1
15 – 19	3	2	1	1	1	1	5	4
20 – 34	5	3	4	2	0	0	9	5
35 – 49	8	6	12	9	0	0	20	15
50 +	1	0	0	0	0	0	1	0
Unknown adult	0	0	0	0	5	1	5	1
Totals	17	11	17	12	9	3	43	26

Results

There are six cases of vertebral fusion observed from Golovin Bay, of which two individuals exhibit block vertebrae that appeared to be congenital based upon those criteria mentioned above. A middle adult female (AMNH-352379A) shows segmentation failure of the second and third cervical vertebrae (Fig. 2). The vertebrae are attached at the inferior articulations of C2 and the superior articulations of C3 and between the laminae of the neural arches. Osteoarthritis is observed on the odontoid process of C2 and the inferior articular surfaces of C3, but it is not severe and is most likely a secondary result of the block vertebrae limiting the movement of the joint. Similarly, a late adolescent classified as a probable male (AMNH-346033) has third and fourth cervical vertebrae that are connected at the left laminae and partially connected on the right laminae as well. No osteoarthritis or osteophyte development was observed in any of the vertebral elements present from this individual. The frequency of vertebrae exhibiting segmentation failure in the Golovin Bay collection is 1.9% (1/54) for C2/C3 and 2.2% (1/46) for C3/C4. Of the four other individuals exhibiting vertebral fusion, three show fusion that was the result of advanced osteophyte development. The fourth individual is classified as a young adult male with probable rheumatoid arthritis, exhibiting fusion of nearly all of the vertebral elements and ribs. This individual was previously examined by Ortner and Putschar (1985: 410), who characterized the condition as 'atypical' rheumatoid arthritis.

Only two individuals in the Nunivak Island collection exhibit fused vertebrae of any kind. A young adult female (AMNH-339228) and an adolescent child (AMNH-339235) each show segmentation failure of the second and third cervical vertebrae (Figures 3 & 4). In each case, C2 and C3 are joined across the neural arches. In the case of the young adult female, the vertebral bodies are also connected, although an intervertebral disc space was retained. There is no indication of osteophytosis in either individual. Moderate arthritis, in the form of porosities on the superior articular facets of C4, is present in the child, and is most likely secondary to the segmentation failure. The frequency of segmentation failure for C2/C3 in the Nunivak Island collection is 4.7% (2/43).

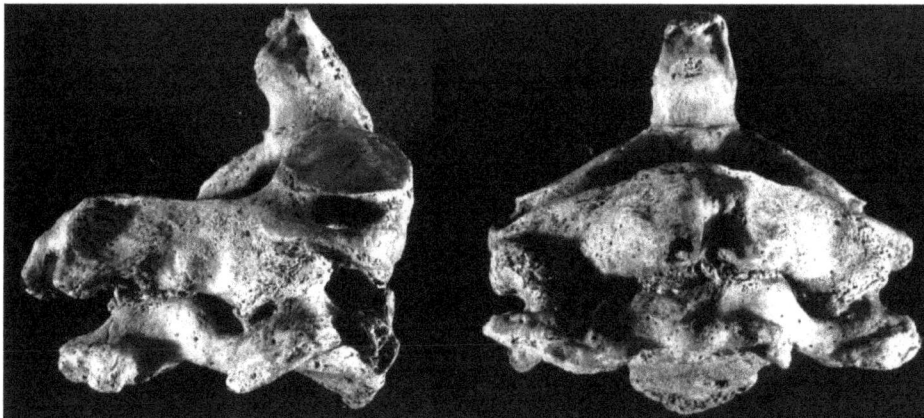

Figure 2. Lateral (left) and posterior (right) views of C2 and C3 from a middle adult female (AMNH-352379A) from Golovin Bay, Alaska.

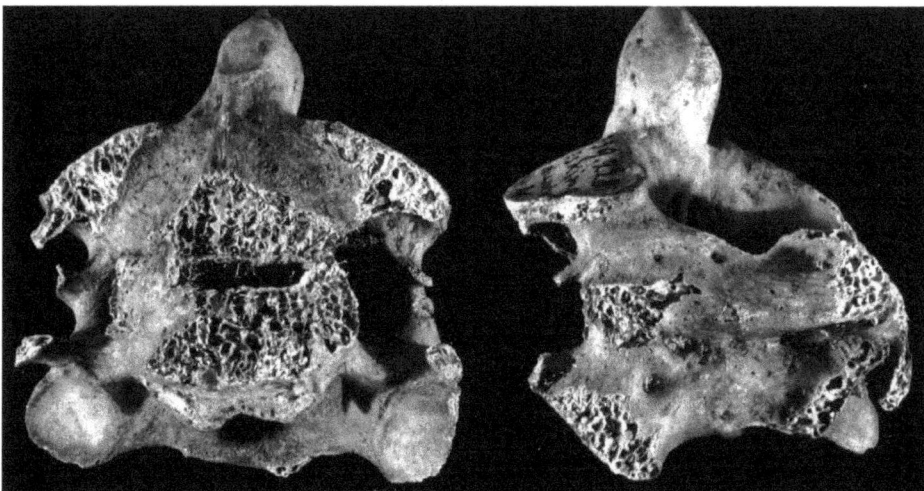

Figure 3. Anterior (left) and lateral (right) views of C2 and C3 from a young adult female (AMNH-339228) from Nunivak Island, Alaska.

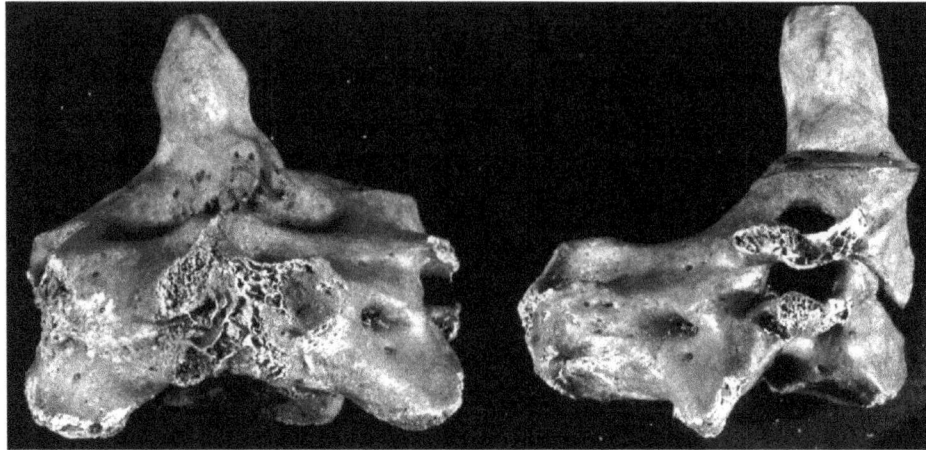

Figure 4. Posterior (left) and lateral (right) views of C2 and C3 from an adolescent child (AMNH-339235) from Nunivak Island, Alaska.

Discussion & Conclusions

The vertebral pathologies described above are most likely related to Klippel-Feil syndrome Type II, characterized by segmentation failure in the developing cervical spine (Spillane *et al.* 1957; Barnes 1994). Genetically, Type II Klippel-Feil syndrome is believed to be an autosomal dominant trait (Barnes 1994), although there is still a great deal of debate on this topic (McKusick 1992; Clarke *et al.* 1998).

Frequencies of occurrence of block vertebrae involving the second and third cervical vertebrae in the two study collections are at 1.9% for Golovin Bay and 4.7% for Nunivak Island. These low frequencies are similar to those reported by Merbs (2004) for two Canadian Arctic populations. He observed segmentation failure for C2 and C3 at frequencies of 0.1% (1/105) and 2.2% (2/92) in collections of Sadlermiut and Thule-Historic Inuit respectively (Merbs 2004). Some researchers have found differential expression by sex. For instance, among a small series of skeletons from northeastern Arizona, attributed to the prehistoric Kayenta Anasazi, five individuals were observed exhibiting congenital fusion of C2 and C3, all of whom were male (Wade, 1981). However, this predominance of males affected differs from the findings of other researchers who observed a greater occurrence of the condition among European females (Gilmour 1941; Gorlin and Pindborg 1964). While one male and one female exhibited segmentation failure in the Golovin Bay collection, no males were affected in the Nunivak Island collection. However, this condition is very rare, making characterization of the trait by sex in either collection difficult. Even so, these observations provide further information regarding a rare congenital condition occurring in two separate Alaskan Native skeletal collections.

Acknowledgments

This research was funded in part by the Smithsonian Institution Office of Repatriation. Special thanks must be made to the people of Golovin Bay and Nunivak Island for allowing this research to be conducted at the University of Alaska. Skeletal analysis assistance was provided by Dr. G. Richard Scott, Steven R. Street, Dr. Susan Steen, Robert Lane, and C. Ryan Colby. Photos were provided courtesy of Dr. G. Richard Scott. Thanks also to Norma Haubenstock and Dr. Michelle Epp for editorial assistance.

References

Aufderheide AC & Rodríguez-Martín C. 1998. *The Cambridge Encyclopedia of Human Paleopathology.* Cambridge University Press: New York.

Barnes E. 1994. *Developmental Defects of the Axial Skeleton in Paleopathology.* University of Colorado Press: Niwot.

Clarke RA, Catalan G, Diwan AD, Kearsley JH. 1998. Heterogeneity in Klippel-Feil syndrome: a new classification. *Pediatric Radiology* 28:967.

Danforth ME, Cook DC, Knick III SG. 1994. The human remains from Carter Ranch Pueblo, Arizona: health in isolation. *American Antiquity* 59:88-101.

Gilmour JR. 1941. The essential identity of the Klippel-Feil syndrome and iniencephaly. *Journal of Pathology and Bacteriology* 53:117-31.

González-Reimers E, Mas-Pascual A, Arnay-De-La-Rosa M, Velasco-Vázquez J, Jiménez-Gómez MC. 2001. Klippel-Feil syndrome in the prehispanic population of El Hierro (Canary Islands). *Annals of the Rheumatic Diseases* 60:173a.

Gorlin RJ & Pindborg JJ. 1964. *Syndromes of the Head and Neck.* McGraw-Hill: New York.

McKusick VA. 1992. *Mendelian Inheritance in Man: Catalogs of Autosomal Dominant, Autosomal Recessive, and X-Linked Phenotypes.* The Johns Hopkins University: Baltimore.

Merbs C, F. 2004. Sagittal clefting of the body and other vertebral developmental errors in Canadian Inuit skeletons. *American Journal of Physical Anthropology* 123:236-49.

Merbs CF & Euler RC. 1985. Atlanto-occipital fusion and spondylolisthesis in an Anasazi skeleton from bright angel ruin, Grand Canyon National Park, Arizona. *American Journal of Physical Anthropology* 67:381-91.

Ortner DJ & Putschar WGJ. 1985. *Identification of the Pathological Conditions in Human Skeletal Remains.* Smithsonian Institution Press: Washington D.C.

Schmorl G & Junghanns H. 1971. *The Human Spine in Health and Disease.* Grune & Stratton: New York.

Speaker S, Kingston D, Mudar KM. 1996. *Inventory and assessment of human remains and associated funerary objects from Nunivak Island, Alaska,* in the National Museum of Natural History. Repatriation Office, National Museum of Natural History, Smithsonian Institution: Washington, D.C.

Spillane JD, Pallis C, Jones AM. 1957. Developmental abnormalities in the region of the foramen magnum. *Brain* 80:11-49.

Verano JW & Urcid J, (eds). 1994. *Physical Anthropology Laboratory Data Manual,* Technical Reports, No. 1. Repatriation Office, National Museum of Natural History, Smithsonian Institution: Washington D.C.

Wade WD. 1981. Klippel-Feil syndrome in a prehistoric population of northern Arizona. In: *Contributions to Physical Anthropology,* 1978-1980. Cybulski JS, (ed). National Museums of Canada: Ottawa.

THE SPECIFICITY OF PALAEOPATHOLOGICAL DIAGNOSIS: A CASE OF BILATERAL SCAPHOLUNATE ADVANCED COLLAPSE IN A ROMANO-BRITISH SKELETON FROM ANCASTER

A.M. Roberts & K. Robson Brown

Abstract

Palaeopathology aims to identify and record the incidence of specific pathologies in archaeological human remains. Palaeopathological diagnosis is most reliable when diagnostic criteria are linked as closely as possible to those used in modern clinical diagnosis. It should be possible to translate specific clinical diagnoses based primarily on radiographic changes into similarly specific palaeopathological diagnoses. Clinically, scapholunate advanced collapse (SLAC) is the most common pattern of degenerative joint disease of the wrist, involving a progressive destruction of the radioscaphoid and then the capitolunate joint. There is only one report of SLAC wrist in the palaeopathological literature. In this paper, we report on another ancient case of bilateral SLAC wrists, found in a Roman skeleton from Ancaster, Lincolnshire. Osteoarthritis (OA) was diagnosed according to accepted palaeopathological criteria: principally the presence of eburnation on a joint surface. Eburnation was found at the articular surfaces of the wrist joint and numerous intercarpal joints bilaterally. The pattern of joints affected matched modern clinical descriptions of SLAC wrist. Radiographic changes characteristic of OA were identifiable at the wrist joint, but not at the intercarpal joints. This case proves that SLAC wrist is an extremely specific diagnosis that nonetheless may be applied to human skeletal remains. The discrepancy between the observational and radiographic findings highlights the problems encountered when attempting to compare disease in archaeological versus modern populations.

Keywords: Palaeopathology; degenerative joint disease; osteoarthritis; wrist; SLAC; Roman

Introduction

Palaeopathology as a discipline aims to identify and record the incidence of specific pathologies in archaeological human remains, an endeavour which can provide data on the impact of disease and trauma on human populations throughout time. As such, it supplements both archaeological enquiry, by providing information on the health of past populations, and medical research, by revealing evidence about the natural history of diseases by applying modern methods of diagnosis to past populations (Larsen 1997; Miller *et al.* 1996). Diagnosis in human skeletal remains is limited by the material available, by the unpredictability of disease and the uncertainty of diagnosis. Palaeopathological diagnosis will be most reliable when diagnostic criteria are linked as closely as possible to those used in modern clinical diagnosis, and with the use of investigative tools such as radiography and analytical tools such as differential diagnosis (Miller *et al.* 1996; Rogers & Waldron 1995).

Although many clinical diagnoses include signs and symptoms which preclude a directly equivalent palaeopathological diagnosis, others may be based primarily on directly visualised changes, such as those seen at arthroscopy, or radiographic changes which may also be seen in dry bones. In the latter, it should be possible to make equivalent palaeopathological diagnoses, as specific as those made in clinical cases.

In this paper, we describe a case of scapholunate advanced collapse (SLAC) found in a Roman skeleton from Ancaster, Lincolnshire. SLAC wrist represents a specific category of degenerative joint disease (DJD) at the wrist, and is the most common pattern of wrist DJD seen clinically, involving destruction of the radioscaphoid and then the capitolunate joint (Watson & Ballet 1984; Watson & Ryu 1986). Despite being a common modern clinical finding, the only palaeopathological evidence for a SLAC wrist to date is a single case of bilateral SLAC wrist in a 7000 year old prehistoric skeleton from Hassi-el-Abiod in the Sahara (Masmejean *et al.* 1997). As SLAC is primarily diagnosed from radiographic signs of DJD in the wrist, it is a good candidate for a palaeopathological entity, as diagnosed by inspection of joint surfaces and radiography in dry bones.

Materials and Methods

Skeleton ANC 01 217 was discovered during a Time Team (Channel 4) excavation in Ancaster, Lincolnshire, in 2001. The stone cist containing the skeleton was found in a trench dug to investigate the extent of the Roman cemetery, to the west of the previously identified Roman town. ANC 01 217 was transported to the Department of Anatomy at the University of Bristol, where the bones were washed and assessed for an osteological report. An inventory of the bones was made, the completeness of the skeleton estimated, and the state of preservation and fragmentation of the bones was also recorded.

The material was aged and sexed according to standard anthropological techniques (Brothwell 1981; Buikstra & Ubelaker 1994; Mays 1998; Cox & Mays 2000). Stature was calculated from long bone measurements, using Trotter's regression formulae (Trotter 1970). Selected metrics and non-metrics were recorded, following the recommendations of the British Association for Biological Anthropology and Osteoarchaeology and the Institute of Field Archaeologists (Brickley & McKinley 2004). Signs of pathology were recorded, and differential diagnoses, together with the most probable diagnosis, offered.

The signs of DJD recorded on the skeleton followed the diagnostic criteria published by Dr Juliet Rogers and her clinical colleagues in Bristol (Rogers *et al.* 1987). These criteria were themselves closely related to the radiographic criteria for the diagnosis of DJD *in vivo*: a narrowed joint space, subchondral sclerosis, subchondral cysts and the presence of marginal osteophytes (Kellgren & Lawrence, 1957; McAlindon & Dieppe 1989). Although the joint space cannot be assessed in skeletal remains, the other radiographic criteria may be successfully applied if the bones are intact. According to Rogers' criteria, eburnation is the pathognomonic macroscopic sign of DJD, and represents the end-stage of degenerative joint disease, where articular cartilage has been lost and subchondral bone exposed. In the absence of eburnation, Rogers' criteria also allow a diagnosis of DJD to be made if two of the following macroscopic signs are present: osteophytes at the margin of the joint, pitting on the joint surface, and alteration of the bony contour (Rogers *et al.* 1987). However, whereas the single criterion of eburnation is universally accepted as pathognomonic of DJD, these other criteria are subject to some controversy. Some researchers reject the proposition that a combination of osteophytes and pitting can be diagnostic of DJD, as each of these features may exist independently, in the absence of DJD (Rothschild 1997; Aufderheide & Rodriguez-Martin 1998). Accordingly, in this report, we made a firm diagnosis of DJD in the presence of eburnation only, although other findings, such as pitting, are also described and illustrated with photographs.

A plain radiograph was made of the wrist and hand bones by placing the bones in contact with the radiographic plate, in the anatomical position, and positioning the x-ray tube 102cm from the plate. An exposure of 45kV, 40mA for 0.45s was used.

Results

Skeleton ANC 01 217 was well preserved and almost complete. The bones were of a light colour, and the majority were intact. The osteological analysis indicated that it belonged to an elderly but robust old adult (50+ years) male, 164.95 +/- 2.99cm tall.

Macroscopic description of lesions in the right wrist and hand

The right distal radius was heavily pitted and eburnated over the area of the articular surface for the scaphoid (Figure 1, RDR), with a reduction in the joint surface of about 3mm in depth, compared with the relatively unaffected lunate facet. There was slight lipping along the dorsal and palmar margins of the facet for the lunate; a small, raised and pitted area at the lateral, palmar margin of facet corresponded with a similarly pitted area on the lunate. The distal ulna was unaffected.

Seven of the eight carpals were present; only the trapezium was missing.

The scaphoid was heavily pitted and eburnated proximally at its facet for the radius (Figure 1, RSP), with pitting extending onto the dorsal surface (Figure 1, RSD); the medial margin of the bone had been worn to a sharp, thin edge. The facet for the capitate was largely unaffected, with only a small, raised area of pitting at the dorsal margin, and the facets for the trapezium and trapezoid were unaffected.

There was an eburnated groove on the lunate to fit the edge of the eburnated area on the distal radius (Figure 1, RLP); the rest of the proximal joint surface of the lunate was mostly smooth with a small area of pitting (about 3mm in diameter) on the palmar margin of the lip lateral to the groove. The lateral facet for the scaphoid was pitted inferiorly and eburnated along the dorsoinferior margin, and along the margin with the capitate facet (Figure 1, RLR). On assembling the bones, it was demonstrated that the capitate had dropped proximally, lateral to the lunate. On the facet of the lunate originally intended for articulation with the capitate, there was an area of pitting and eburnation measuring approximately 6mm^2 and lying adjacent to the dorsal margin of the facet (Figure 1, RLD). The medial joint surface of the lunate (for hamate and triquetral) was preserved.

On the triquetral, there was very slight lipping along the dorsal and palmar margins of the facet for the lunate. The hamate facet was largely smooth, but with a semicircular area (~7mm diameter) of pitting and subtle new bone formation adjacent to the palmar edge (corresponding with an area of pitting on the triquetral facet of the hamate). The facet for the pisiform was largely unaffected, with only a very small area of new bone formation at the superomedial margin.

The facet for the triquetral on the pisiform bone was unaffected.

On the hamate, the medial three quarters of the facet intended for the triquetral was largely unaffected, with a small area of pitting along the palmar margin corresponding with that on the triquetral. The lateral quarter of this facet, however, was eburnated (Figure 1, RHU), with additional pitting in the palmar part of this eburnated area, which extended into the lunate facet. This corresponded with the wear on the lunate at its facet for the capitate, which had been brought into contact with the hamate as the scaphoid had worn down, allowing the capitate to collapse proximally into the gap lateral to the lunate. The facet for the capitate was, perhaps somewhat surprisingly, spared, as was that for the fifth metacarpal.

On the capitate, the facet for the hamate was relatively unaffected. There was a 2mm wide line of eburnation at the angle between the hamate and lunate facets (Figure 1, RCU), where the capitate had ground against the lateral edge of the lunate as it collapsed. The scaphoid facet was relatively unaffected, as were the facets for the trapezoid, and those for the metacarpals.

There was no evidence of degenerative changes on the articular facets of the trapezoid, but some *post mortem* damage on the facet for the capitate.

All five right metacarpals were preserved, as were all right proximal and middle phalanges. There were two

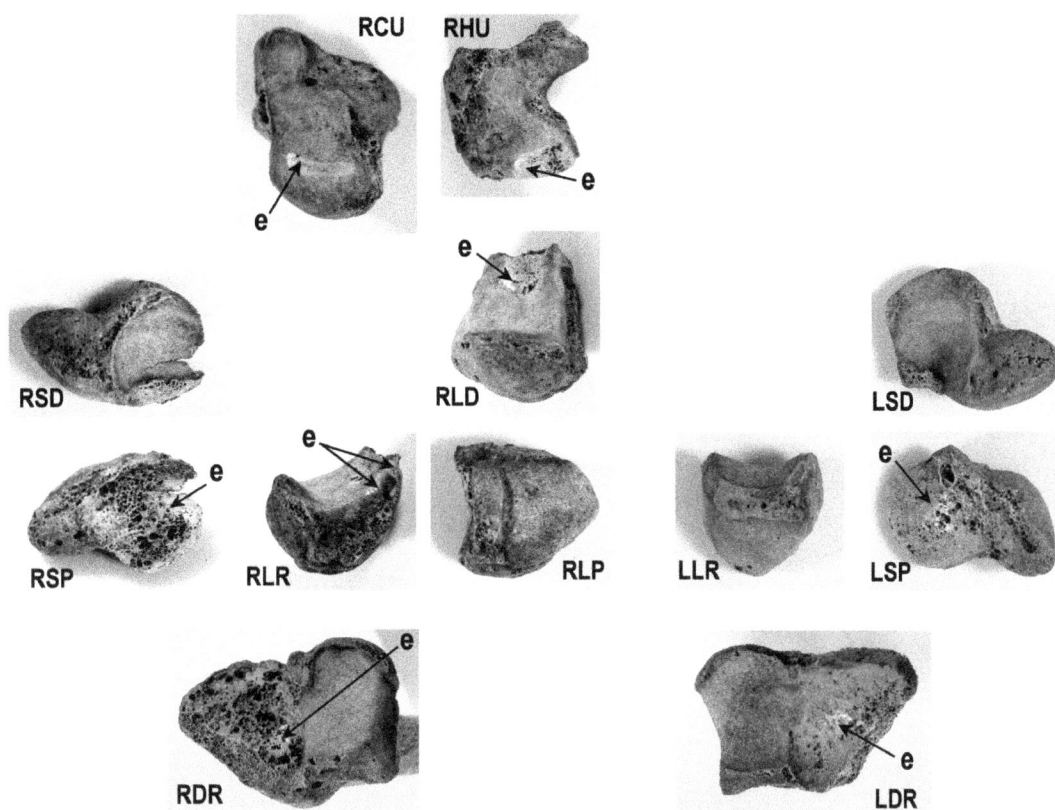

Figure 1: Photographs showing eburnation on the radii and carpals of skeleton ANC 01 217.

RDR: right distal radius; RSP: right scaphoid, proximal aspect; RSD: right scaphoid, distal aspect; RLR: right lunate, radial aspect; RLP: right lunate, proximal apsect; RLD: right lunate, distal aspect; RHU: right hamate, ulnar aspect; RCU: right capitate, ulnar aspect; LDR: left distal radius; LLR: left lunate, radial aspect; LSP: left scaphoid, proximal aspect; LSD: left scaphoid, distal aspect. Areas of eburnation labelled 'e'.

distal phalanges: the 1st and one other. There was severe pitting and erosion at the second (index finger) proximal interphalangeal joint (Figure 2).

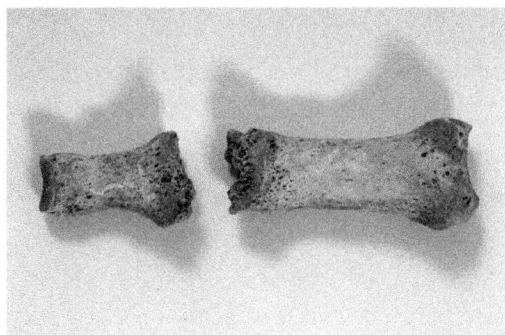

Figure 2: Photograph of the proximal and middle phalanges of the right second digit (index finger) showing erosion of the distal epiphysis of the proximal phalanx and of the proximal epiphysis of the middle phalanx (ie: at the proximal interphalangeal joint).

Macroscopic description of lesions in the left wrist and hand

The left distal radius presented an area of eburnation with moderate pitting over the articular surface for the scaphoid, but with negligible reduction in the level of the joint surface (Figure 1, LDR). The lunate facet was largely unaffected, with only very slight lipping at the dorsal and palmar margins.

All eight left carpals were present.

The scaphoid was eburnated and pitted on the radial facet (Figure 1, LSP), although not as severely as the right. The facet for the capitate was largely unaffected, with a small, raised area of pitting at the dorsal margin (Figure 1, LSD), similar to that on the right scaphoid. The facets for the trapezium and trapezoid were unaffected.

The radial articular surface of the lunate was mostly unaffected; there was some postmortem damage to the ulnar edge of this surface. There was new bone growth at the area between the radial and scaphoid facets (Figure 1, LLR). The facets for the scaphoid, capitate and hamate were themselves unaffected. There was very slight osteophytic lipping at the dorsal margin of the triquetral facet.

The pisiform appeared to be waisted, which may be evidence of a healed fracture or a developmental variation. The facet for the triquetral was unaffected.

The triquetral, hamate, capitate, trapezoid and trapezium showed no evidence of joint disease.

All five left metacarpals were present and intact, as were the five proximal and four middle phalanges. There were three distal phalanges, including the first. There was no evidence of joint disease at the carpometacarpal, metacarpophalangeal or interphalangeal joints.

Radiographic description of lesions in the right wrist and hand (Figure 3a)
Subchondral sclerosis was present at the distal end of the right radius, together with prominent subchondral cysts, particularly evident in the radial styloid process and beneath the facet for the lunate on the ulnar side of the distal radius. No changes were visible in the distal ulna.

The scaphoid was narrowed to a thin point, with sclerosis at both the proximal (radial) and distal (capitate) articular surfaces. A large subchondral cyst was present in the body of the bone.

On the proximal surface of the lunate, the groove (about 2mm in depth) was clearly evident on the radiograph; there were no subchondral cysts or sclerosis underlying this surface. A dense line of sclerosis on the distal surface (for the capitate) extended onto the lateral surface (originally articulating with the scaphoid, but later coming into contact with the collapsed capitate). Subchondral cysts lay beneath these articular surfaces. No changes were observed on the ulnar side of the lunate.

On the capitate, a small patch of dense subchondral sclerosis was evident at the inferomedial angle (between the hamate and lunate facets). Subchondral osteopaenic areas were apparent at the head of the capitate and on its lateral border, although these could be due to *post mortem* taphonomic changes rather than *ante mortem* pathological processes.

On the hamate, a possible area of subchondral sclerosis was evident beneath the articular surface for the fifth metacarpal.

No obvious radiographic changes were observed in the triquetral, pisiform or trapezoid.

The metacarpals were unaffected. The second (index finger) proximal interphalangeal joint was eroded, with the articular surface destroyed on both sides. There was no sclerosis at this joint.

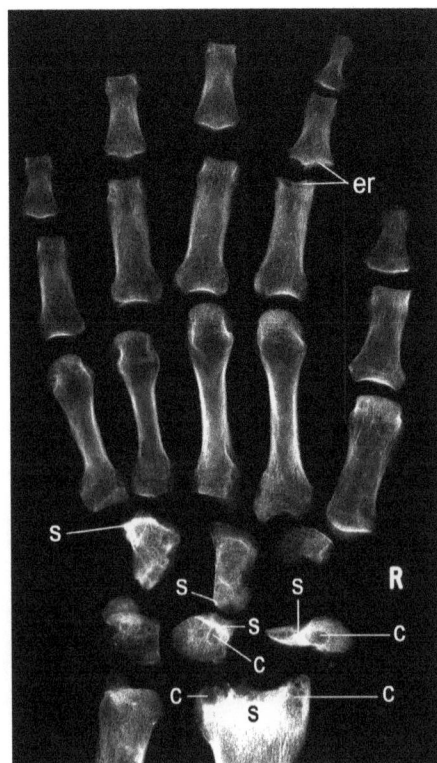

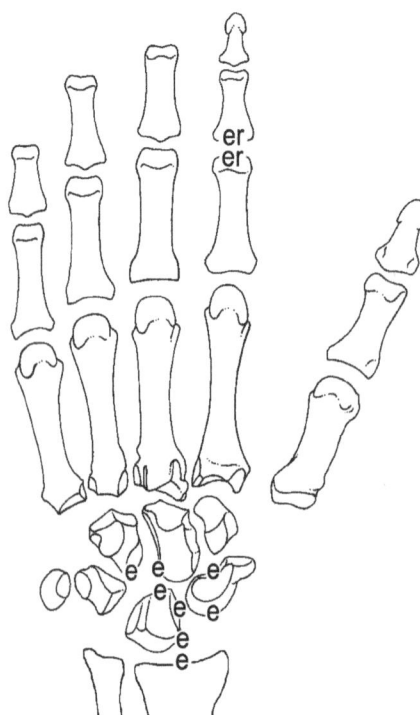

3a: Right wrist

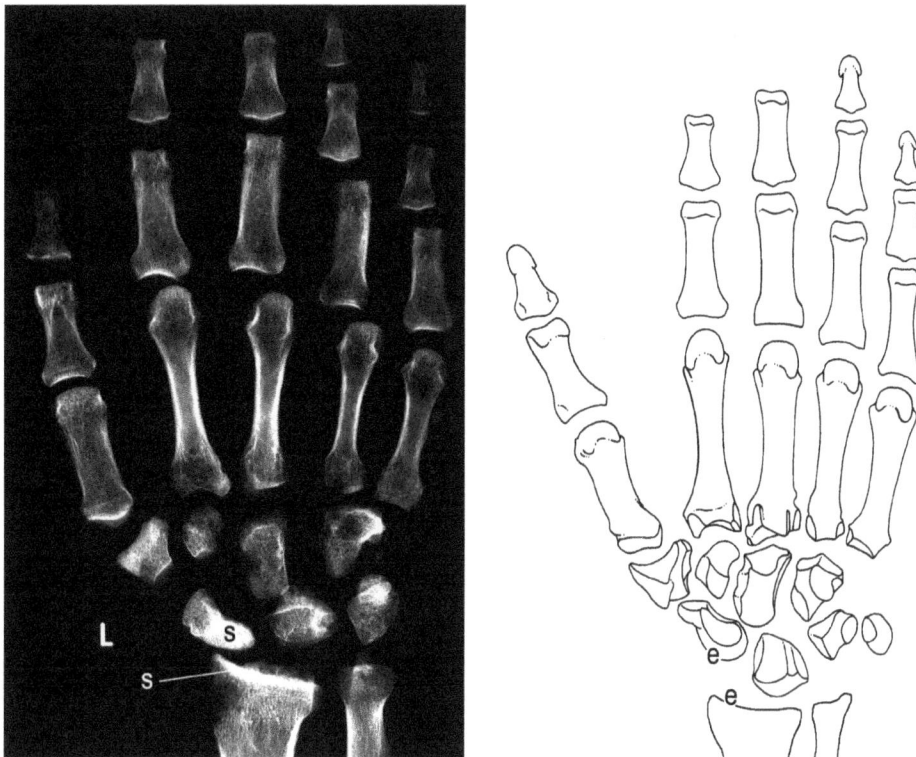

3b: Left wrist

Figure 3: Macroscopic and radiographic signs of DJD in the wrists of ANC 01 217
Radiographic changes are labelled s (subchondral sclerosis), c (subchondral cyst) and er (erosion) in the radiographs
on the left; the macroscopic changes are labelled on the diagrams on the right: e (eburnation) and er (erosion)

Radiographic description of lesions in the left wrist and hand (Figure 3b)
A dense band of subchondral sclerosis was present at the distal articular surface of the left radius. A prominent osteopaenic area occupied the lateral margin of the distal radius, proximal to the radial styloid process. This area was not well circumscribed, so could be taphonomic rather than pathological.

The medial half of the scaphoid was sclerosed; no subchondral cysts were visible.

No obvious radiographic changes were observed in the other carpals, except for an osteopaenic area on the lateral side of the capitate, which could be taphonomic rather than pathological.

No radiographic changes were seen in the metacarpals or phalanges.

The occurrence of macroscopic eburnation and the radiographic signs of DJD in the wrists of ANC 01 217 are combined and summarised in table 1, and illustrated in Figure 3.

Other pathology
Several other pathological lesions were present elsewhere in the skeleton, and these are summarised in Figure 4.

Discussion
The changes seen on macroscopic observation of the bones, and on radiography, are mostly consistent with each other (see Table 1). In a large study of the radiographic-anatomic correlation of the diagnosis of arthritis, where radiographic findings were compared with actual cartilage loss at joints in 393 wrists, radiography was found to be a reliable indicator of arthritis at the radioscaphoid joint, but not at other carpal joints (Peh *et al.* 1999). In this study, findings of eburnation on the dry bones were matched by findings of subchondral sclerosis and cysts on the right radius, scaphoid and lunate. The small area of eburnation on the right triquetral was not accompanied by any radiographic signs. The eburnation on the left radius and scaphoid were accompanied by subchondral sclerosis but no cysts. The radiography of skeletal material is not directly comparable with clinical radiographs where the bones are united by soft tissues. The absence of soft tissues shadows may make visualisation of some signs easier (eg: sclerosis, subchondral cysts), while other signs such as joint space narrowing are lost to history.

The pattern of DJD, both macroscopic and radiographic, in both wrists of the Ancaster 01 217 skeleton, is consistent with a diagnosis of scapholunate advanced collapse (SLAC). This diagnosis may be further refined into stages. SLAC stage I represents the initial phase of

TABLE 1: MACROSCOPIC EBURNATION AND RADIOGRAPHIC SIGNS OF DJD IN THE WRISTS OF ANC 01 217

Bone	Articular facet	Right wrist				Left wrist			
		E	S	C	O	E	S	C	O
Distal radius	Scaphoid	1	1	1	0	1	1	0	0
	Lunate	0	1	1	0	0	1	0	0
Scaphoid	Radial	1	1	1	0	1	1	0	0
	lunate	1	1			0	1		
	Capitate	0	1			0	1		
	Trapezium, trapezoid	0	0			0	0		
Lunate	Radial	1	0	1	0	0	0	0	0
	Scaphoid	1	1			0	0		
	Triquetral	0	0			0	0		
						0	0		
	Hamate	1	0			0	0		
	Capitate	1	1			0	0		
Triquetral	Hamate	0	0	0	0	0	0	0	0
	Lunate	0	0			0	0		
	Pisiform	0	0			0	0		
Pisiform	Triquetral	0	0	0	0	0	0	0	0
Trapezium	Trapezoid	#				0	0	0	0
	scaphoid					0	0		
	Metacarpal I					0	0		
	Metacarpal II					0	0		
Trapezoid	Trapezium	0	0	0	0	0	0	0	0
	Scaphoid	0	0			0	0		
	Capitate	#	0			0	0		
	2nd metacarpal	0	0			0	0		
Capitate	Lunate	1	1	0	0	0	0	0	0
	Scaphoid	0	0			0	0		
	trapezoid	0	0			0	0		
	hamate	0	1			0	0		
	2nd metacarpal	0	0			0	0		
	3rd metacarpal	0	0			0	0		
Hamate	Capitate	0	0	0	0	0	0	0	0
	Lunate	1	0			0	0		
	Triquetral	1	0			0	0		
	4th metacarpal	0	0			0	0		
	5th metacarpal	0	0			0	0		

E = macroscopic eburnation
S = radiographic subchondral sclerosis
C = radiographic subchondral cysts
O = radiographic marginal osteophyte
# = missing or damaged
Signs are scored as present (1) or absent (0)

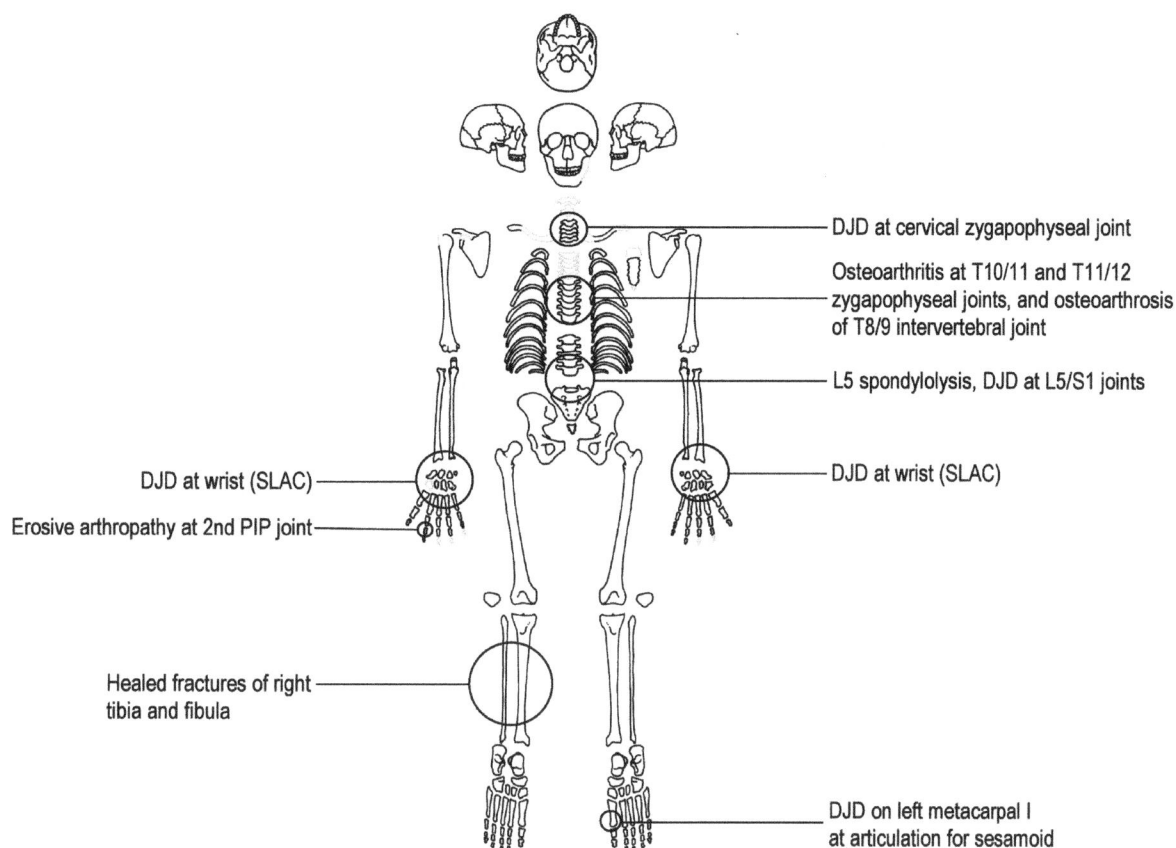

Fig 4: Visual record of skeleton ANC 01 217 with summary of pathology

this degenerative process, where cartilage is lost at the distal pole of the scaphoid and the radial styloid. SLAC stage II also involves cartilage loss at the proximal pole of the scaphoid and the entire scaphoid fossa of the radius. SLAC stage IIIa marks progression to cartilage loss at the capitolunate joint, and stage IIIb represents the collapse of the capitate into the opening gap between the scaphoid and the lunate, with cartilage loss at the scaphocapitate joint. At all of these stages, the radiolunate joint remains unscathed. The radiographic diagnosis and staging of SLAC wrist in modern clinical cases therefore depends on assessing the progression of degenerative changes seen at the radioscaphoid and intercarpal joints, as evidenced by loss of joint space and sclerosis (Cooney et al. 1998; Metz et al. 1997; Stabler et al. 1997; Tang et al. 2002).

From the pattern of DJD at the radiocarpal and intercarpal joints of Ancaster 01 217 skeleton, and the position that the assembled bones assume, the diagnosis may be refined to SLAC stage II in the left wrist (with eburnation over the entire radioscaphoid joint, but no evidence of cartilage loss at the capitolunate joint), and SLAC stage IIIb in the right wrist (with eburnation at the capitolunate joint, and dislocation of the capitate into the gap between the scaphoid and lunate).

SLAC wrist is a post-traumatic condition: the most common cause is traumatic injury to the scapholunate

interosseous ligament, which often occurs as a result of a fall onto an extended wrist (Meldon & Hargarten 1995; Metz et al. 1997; Stabler et al. 1997; Cooney et al. 1998; Crisco et al. 2003). A similar pattern of DJD in the wrist may develop secondary to non-union of a fractured scaphoid (ie: scaphoid non-union advanced collapse: SNAC). Although definitive SLAC wrist is a post-traumatic condition, similar patterns of wrist DJD may also arise secondary to primary degenerative arthritis and primary ligamentous instabilities (such as capitolunate degeneration or midcarpal instability). Occasionally inflammatory arthritis (including rheumatoid arthritis, gout and calcium pyrophosphate deposition disease), osteochondritis (eg: Kienbock disease), neuropathic diseases, and amyloidosis may cause a similar pattern of malalignment and secondary DJD of the carpal bones (Chen et al., 1990; Stabler et al., 1997). In ANC 01 217, the pitting and erosion at the interphalangeal joint in the right hand of ANC 217 (Figure 2) represents erosive arthropathy at this joint. The differential diagnoses for such erosive arthropathy include rheumatoid arthritis (RA), spondyloarthropathy and infective arthritis. Although RA may occasionally cause a similar pattern of arthritis to SLAC, RA an unlikely diagnosis for any of the joints in the wrist and hand, as the bones are not osteopaenic, and in particular, the lesion in the interphalangeal joint is isolated and asymmetrical (Blondiau et al. 1997). Asymmetrical and oligoarticular involvement is typical of spondyloarthropathy, although in the absence of any sacroiliac disease this is also an insupportable diagnosis. The erosive arthropathy at this single interphalangeal joint is therefore best explained as an infective monoarthritis (Inoue *et al.,* 1999). As there is no evidence for rheumatoid arthritis, or any other systemic factor predisposing to this particular pattern of wrist DJD, the most likely explanation for this bilateral condition in the Ancaster 217 skeleton is indeed the post-traumatic DJD recognised in the modern clinical setting as SLAC.

This investigation demonstrates that SLAC is identifiable as a palaeopathological entity, although the discrepancy between the observational and radiographic findings highlights the problems encountered when attempting to compare disease in archaeological versus modern populations. Although SLAC is a common modern clinical finding, it was only specifically identified during the last thirty years, and this is the second reported case of SLAC in the palaeopathological literature. It is unlikely that SLAC was rare in the past, and more likely that it has not been specifically diagnosed, but instead has been identified more broadly as 'wrist osteoarthritis'. The medical literature reflects a recent expansion of knowledge concerning the diagnosis, classification and causes of wrist DJD (Cooney *et al.* 1998), and this presents a challenge to palaeopathologists. If manifestations and frequencies of disease in past populations are to be compared with those in modern populations, palaeopathological diagnosis should attempt to be as specific as current medical diagnosis. Following the clinical model, differential diagnoses should be offered. All observations should be carefully recorded; this is extremely important as it allows the information to be used in comparative studies, even if different diagnostic criteria have been used in the original analyses. It will also allow future workers to make retrospective diagnoses as medical diagnosis evolves. Palaeopathologists should also use similar techniques (such as radiography) wherever possible, and should base their diagnostic criteria on those used in modern clinical diagnosis.

Acknowledgements

We would like to acknowledge the following people for their help on this project:
Kate Edwards (Time Team, Videotext) for permitting the study of this skeleton; Nicola Latham (Department of Anatomy, University of Bristol) for the excellent radiography of the wrist and hand bones; Dave Newbury (Department of Anatomy, University of Bristol) for the beautiful photographs of the bones.

References

Aufderheide AC, Rodriguez-Martin C. 1998. *The Cambridge Encyclopaedia of Human Palaeopathology.* Cambridge University Press: Cambridge.

Blondiau J, Cotton A, Fontaine C, Hanni C, Bera A, Flipo R. 1997. Two Roman and Medieval Cases of Symmetrical Erosive Polyarthropathy from Normandy: Anatomicopathological and Radiological Evidence for Rheumatoid Arthritis. *International Journal of Osteoarchaeology.* 7: 451-466

Brickley M & McKinley JI. 2004. *Guidelines to the Standards for Recording Human Remains.* IFA Paper No. 7: Reading.

Brothwell DR. 1981. *Digging Up Bones* (3rd edition). Oxford University Press/British Museum (Natural History): Oxford.

Buikstra JE & Ubelaker DH (eds.). 1994. *Standards for data collection from human skeletal remains.* Arkansas Archaeological Survey: Arkansas

Chen C, Chandnani VP, Kang HS, Resnick D, Sartoris DJ, Haller J. 1990. Scapholunate Advanced Collapse: A Common Wrist Abnormality in Calcium Pyrophosphate Dihydrate Crystal Deposition Disease. *Radiology.* 177: 459-461.

Cooney WP, DeBartolo T, Wood MB. 1998. Post-traumatic arthritis of the wrist. In *The Wrist.*, Cooney WP, Linscheid RL, Dobyns JH (eds.). Mosby.

Cox M & Mays S. 2000. *Human Osteology.* Greenwich Medical Media: London

Crisco JJ, Pike S, Hulsizer-Galvin DL, Akelman E, Weiss AC, Wolfe SW. 2003. Carpal Bone Postures and Motions Are Abnormal in Both Wrists of Patients with Unilateral Scapholunate Interosseous Ligament Tears. *The Journal of Hand Surgery.* 28: 926-937.

Inoue K, Hukuda S, Nakai M, Katayama K, Huang J. 1999. Erosive Polyarthritis in Ancient Japanese Skeletons: A Possible Case of Rheumatoid Arthritis. *International Journal of Osteoarchaeology.* 9: 1-7

Kellgren JH and Lawrence JS. 1957. Radiological Assessment of Osteoarthritis. *Annals of the Rheumatic Diseases.* 16: 494-501.

Larsen C S (1997) *Bioarchaeology.* Cambridge University Press, Cambridge.

McAlindon T & Dieppe P. 1989. Osteoarthritis: Definitions and Criteria. *Annals of the Rheumatic Diseases.* 48: 531-532.

Masmejean E, Dutour O, Touam C and Oberlin C. 1997. Bilateral SLAC (scapholunate advanced collapse) wrist: an unusual entity. Apropos of a 7000-year-old prehistoric case. *Annales de chirurgie de la main et du membre superieur.* 16: 207-214

Mays S. 1998. *The Archaeology of Human Bones.* Routledge: London.

Meldon SW & Hargarten SW. 1995. Ligamentous Injuries of the Wrist. *The Journal of Emergency Medicine.* 13: 217-225.

Metz VM, Metz-Schimmerl SM, Yin Y. 1997. Ligamentous instabilities of the wrist. *European Journal of Radiology.* 25: 104-111

Miller ERB & Ortner DJ (1996) Accuracy in dry bone diagnosis: a comment on palaeopathological methods. *International Journal of Osteoarchaeology* 6(3): 221-229.

Peh WCG, Patterson RM, Viegas SF, Hokanson JA, Gilula LA. 1999. Radiographic-anatomic correlations at different wrist articulations. *The Journal of Hand Surgery* 24A: 777-780.

Rogers J, Waldron T, Dieppe P, Watt I. 1987. Arthropathies in palaeopathology: The basis of classification according to most probable cause. *Journal of Archaeological Science.* 14: 179-193.

Rothschild BM. 1997. Porosity: A Curiosity without Diagnostic Significance. *American Journal of Physical Anthropology.* 104: 529-533.

Stabler A, Heuck A, Reiser M. 1997. Imaging of the hand: degeneration, impingement and overuse. *European Journal of Radiology.* 25: 118-128.

Tang JB, Ryu J, Omokawa S, Wearden S. 2002. Wrist kinetics after scapholunate dissociation: the effect of scapholunate interosseous ligament injury and persistent scapholunate gaps. *Journal of Orthopaedic Research.* 20: 215-221.

Trotter M. 1970. Estimation of stature from intact long limb bones. In *Personal identification in mass disasters.*, Stewart TD (ed.). National Museum of Natural History, Smithsonian Institution: Washington DC

Watson HK & Ballet FL. 1984. The SLAC wrist: scapholunate advanced collapse pattern of degenerative arthritis. *Journal of Hand Surgery* 9: 358-365

Watson HK & Ryu J. 1986. Evolution of arthritis of the wrist. *Clinical Orthopaedics* 202: 57-67.

A ZOOARCHAEOLOGICAL CONTRIBUTION TO BIOLOGICAL ANTHROPOLOGY: WORKING TOWARDS A BETTER UNDERSTANDING OF CUT MARKS AND BUTCHERY

K. Seetah

Abstract

Due to the obvious constraints involved with working on human osseous material many advances in experimental aspects of biological anthropology have made use of research from the zooarchaeological analysis of faunal remains. Cut marks represent unequivocal evidence for human activity on animal bones, making the analysis of butchered remains an important part of zooarchaeological research. However, understanding the technical aspects of dismemberment is crucial to interpreting activity from assemblages where cannibalism has occurred. As faunal butchery has played an important part in developing an understanding of how humans may have used implements on other humans, it is equally important for those involved in human osteoarchaeology to also have an appreciation of how orientation, implement use and fundamental carcass dismemberment principles may affect how and why a particular process was carried out. This paper will highlight aspects of butchery and carcass dismemberment that may assist with interpretation and appraisal of cut marks on human remains, for example in distinguishing dismemberment for nutrition as opposed to cut marks made during ritual/sacrificial practices.

Keywords: butchery; zooarchaeology; osteoarchaeology; human remains; cut-marks

Cut mark research in context

The analysis of cut marks has become a vital means of achieving interpretive data from osseous remains. No other analytical tool indicates the unequivocal intent of humans to carry out a specific process, in this case, either as part of a procurement strategy or for ritual practice. This paper will explore some of the advances made through zooarchaeological research of processing marks, highlighting how these developments might find application as part of the identification and interpretation of cut marks noted on human remains within an archaeological context.

The use of zooarchaeology to assist with interpretations from human remains is not a novelty. The specific constraints placed on physical and forensic anthropologists have often led to a dependence on faunal research to assist where experimentation on human material would be prohibited. Thompson (2002, 2004), for example, studied the changes that occur to bone during various stages of burning. The research demonstrated that particular caution needed to be exercised when dealing with sexing of cremated material; throughout the experimental phase of the study there was a dependence on animal bone. While this example demonstrates the oft incurred reliance on animal bone as

a raw material, other research has borrowed comprehensively from zooarchaeology. The taphonomic history of the human bone assemblage from Sima de los Huesos, Atepuerca, was analysed using terminology and models following Binford (1981) in order to distinguish between scavenging / carnivore activity and trampling damage (Andrews and Fernandez-Jalvo 1997). Studies of cannibalistic activities at Mancos Canyon, Colorado incorporated a broad appropriation of zooarchaeological theory and method (White 1992). In this study White undertook an appraisal of the human remains, incorporating perspectives from Lyman (1978, 1987) and Binford (1981), in an attempt to clarify what caused the particular pattern of cut marks and bone fragmentation.

The present article continues this trend by presenting aspects of ongoing zooarchaeological research into butchery and processing practices in order to outline how this might be used to better understand cannibalism. Of key importance is that an understanding of *process* during disarticulation is imperative and should be axiomatic; a fundamental criterion that merits concise and consistent appraisal.

The importance of cut mark research lies in its inherent dependence on an interdisciplinary approach. A true and accurate appraisal of butchery practice cannot be carried out without a lucid understanding of the tools that have created the mark, or knowledge of the soft tissues that surround the skeletal material. Focusing specifically on the relevance of cut mark analysis to human contexts, three main research areas are evident; the identification of subsistence strategies[1], the clarification of ritual practice and exploring aspects of human conflict.

Integrating Zooarchaeology and Physical Anthropology/Forensic Archaeology

Within a zooarchaeological context cut mark analysis is invariably employed to explore economics, subsistence and/or procurement of meat, despite the fact that it can tell us a great deal about the perceptions and cultural contexts of animals (Seetah 2005). The work presented here has its naissance in research employing modern butchery as a means of understanding aspects of Romano-British meat processing. Thus, it is probably most accurately applied to looking at instances where humans may have been disarticulated as part of a procurement strategy. This *caveat* noted, developing an understanding of the process involved with regard to how the marks may have been inflicted on the skeleton, will certainly be of benefit in any instance where sharp

---

[1] The use of the term 'subsistence' strategies is employed in the loosest sense as there is no intention to imply that humans would ever form the main or dedicated 'meat source'.

implement trauma or ritual activity is suspected. This article is not a comprehensive evaluation of cut marks found on humans in relation to conflict, subsistence or ritual practice; rather it is an appraisal of factors that need to be addressed by the human osteoarchaeologist beyond the mere occurrence, level of fragmentation and location of where marks appear on the skeleton. It is aimed at providing a means of clarifying what cut marks on human bone signify and the types of interpretation they can lead to. This information can then be used to better understand aspects of culture, violence and the use of human flesh as a source of meat.

Conflict & Violence
Conflict between humans occurs in many forms. The inventiveness employed to create and utilize implements to inflict trauma on other humans seems to be limited only by the imagination of the mind, and the resources available to hand; as Gerssdorf (1517) so clearly illustrates (Figure 1).

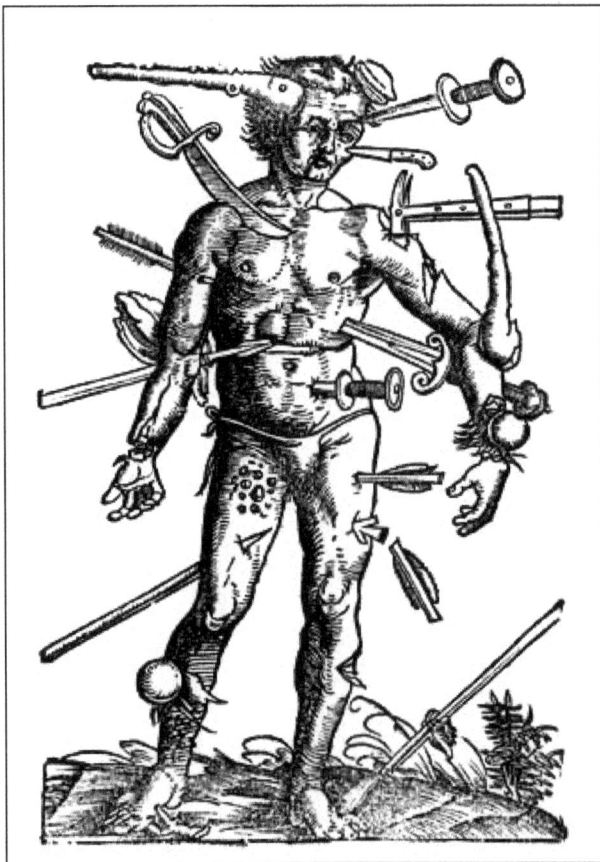

Figure 1. The Wounded Man, from Gerssdorf's Feldbuch der Wundartzney (1517)

The majority of trauma evidenced from human skeletal assemblages can be placed into two main categories; those caused through warfare where violence has occurred on a scale of mass combat, and those caused through inter-personal violence between two or a relatively small number of individuals.

It may seem that cut mark analysis has little to contribute to this particular area, however, this is not the case. Generally there will be a lower occurrence of actual marks than for subsistence, as the intent is likely to be driven by a desire to cause as much injury and/or death as quickly as possible. However, the marks should lend themselves to clearer interpretation compared to a situation where there were multiple marks occurring in a complex matrix (as a result of a combination of procurement practices i.e. skinning, defleshing etc). Furthermore, two key areas do potentially benefit from a better understanding of the cut marks, namely accuracy of identification and recognition of subsequent activity.

Misidentification of marks is a significant problem and one that is particularly relevant where a single traumatic incident is evidenced (Boylston 2000: 359). Larsen (1997: 109) laments the problems of inaccurate identification leading to peculiar inferences relating to conflict and reconstructions of combat. To rectify this situation, taphonomic (and forensic) methodologies should be incorporated into the analysis of cut marks to improve rigour and reliability. Thus, it is clear that actual cut marks need to be clearly differentiated from post-depositional taphonomic processes, including damage created during excavation, in particular the ubiquitous and much maligned trowel marks.

Deciphering subsequent activity after a conflict situation also poses potential problems. It may be the case that after the relatively clear 'trauma' marks are identified, ensuing de-fleshing or disarticulation goes unobserved. In this way, intriguing cultural activity may be obfuscated because of a failure to recognise indicators of less prominent activity. Related to this is the identification of different types of activity that also results in the occurrence of mass traumatic incidences, for example, distinguishing warfare from massacre, genocide (Kennedy 1994) or forms of execution that involve dismemberment. These are likely to elicit marks that are evidently made from 'weapons', but it is essential to differentiate these from marks made in combat. This can be assisted, and indeed achieved, by improving our level of identification and by understanding the stages of activity that took place prior to and during the infliction of the mark.

Ritual & Perception
Interpreting the ritual treatment of human bodies, whether within a burial or sacrificial context, can provide important data for understanding past cultural activity. While it is the case that cut mark analysis of faunal remains invariably leads to conclusions relating to economics, when marks appear on humans there is a tendency to assume a ritual connotation. It must be appreciated that even when subsistence activity is evident, the potential for ritualistic involvement should not go un-noted.

While it is not the intention of this article to delve too deeply into 'ritual activity' *per se*, it is important to relate how aspects of cut mark analysis might be relevant to understanding ritual behaviours, and the perceptions of humans and animals.

One way that cut marks might be linked to ritual activity (and indeed procurement strategies) is to look at whether there is a difference in the way humans have been treated compared to other animals. A key feature of carcass processing is that no matter what species is being dealt with, the overall morphology of the animal will, to a certain extent, dictate how the disarticulation process is carried out. This needs to be evaluated in light of the implements used; however, it is usually the case that sites of 'natural disarticulation' such as joint margins will be the obvious choice for dismembering. A similar disarticulation pattern is likely to be used for all fauna and any deviation from this with any particular species may be indicative of special treatment. Thus, it is possible to observe whether there is a pattern of human processing that mimics 'faunal butchery' and what the implications of this might be.

Generally, within a temperate climatic context, animals being hunted and subsequently processed will usually be cursorial. Within a tropical and subtropical environment the presence of arboreal primates may give clues as to whether, on a purely functional level, there are differences in methods of butchery based on morphology. Intriguing recent research from the Niah Caves, Malaysia, has shown that butchery of primates can follow an almost identical pattern as those employed on other fauna, such as pigs (Rabett & Piper nd.).

One might expect a different means of disarticulation of humans due to the unique morphology and mode of locomotion. If the same butchery techniques are used for animals, particularly primates, but a different range of cut marks are employed for humans, this might indicate a special perceptive view of humans, resulting in associated ritual practice.

How far this potentially extends is not always realised; methods of hunting and subsequent processing are often closely linked to the perceptions of the animal in question (Pluskowski 2005). This aspect of cut mark analysis is particularly important when humans form the 'meat source', as has been demonstrated by ethnographic studies. The Huaorani, an Amazonian hunter-gather group, view their implements either as 'hunting' or 'killing' tools. Blow pipes are used to *hunt* arboreal primates and birds; spears to *kill* the white lipped peccary and humans during conflict. Generally, groups that are not Huaorani are considered to be 'others' and are thought of as cannibalistic as opposed to the Huaorani who are 'true humans'. Within Huaorani society, which is highly dependent on isolationism, the peccary is killed when it comes into the domain of the Huaorani, a situation repeated for people who are not Huaorani. Primates are actively hunted and a great deal is known about their social organisation. This distinction extends into the way tools are made and used as well as how the animals are processed (Rival 1996: 145-164). Although Rival does not discuss how animals or other humans are actually 'butchered', this study does illustrate how perceptions of species, environment, salubriousness and social organisation affect the way processing is carried out.

Crucial to the idea of humans being seen as meat is the image and perception of nature within the society in question (Howell 1996: 127-144). Not surprisingly, not all groups think of themselves as above and beyond nature, as is the case in western philosophical thought and morality (Ingold 1988).While there are numerous detailed accounts of the types of cut marks that might be observed as a result of mortuary practice and other more commonly observed ritual activities (and the relevance of economic factors such as meat quality and how these impact on methods of processing can be appreciated), there is often a failure to link the perceived attitudes of people towards that which they are processing.

Anthropophagy & Subsistence
No appraisal of cut marks on humans is complete without a discussion of cannibalism. Studies have focused on the modifications found on human bone as a result of this activity (Hurlbut 2000) as well as the links with the perceptual and ritual attitudes of those eating their fellow human (Ogilvie & Hilton 2000). Cutting a human and eating the flesh has multiple implications; it can be seen as the ultimate act of humiliation where eating the flesh of an enemy implies an appropriation of their power and strength. Alternatively, within a mortuary context, eating the flesh of a deceased group member may imbue their knowledge and skills to those who participate in the ceremony. This is the area where zooarchaeological research, and in particular the specific research presented here, has the greatest contribution to make to human osteoarchaeology.

What is evident is that cannibalism has been almost universally practiced at one time or another in the majority of geographic regions, with the exception of Europe, within historic times. Despite its long history and wide spatial occurrence, clear identification of cannibalism is not easily accomplished (Wells 1964: 138-9).

Understandably, faunal butchery and disarticulation is easier to accept, particularly due to its considerably higher level of occurrence. However, one should not lose sight of the notion that under certain circumstances humans may be treated, through disarticulation, in the same manner as other animals. This extends to secondary processing through marrow extraction, cooking and processing of human bone for tools. While the potential for cut marks to indicate a perceptual difference between how humans and animals are viewed has been highlighted, the situation may also operate in reverse. Some societies may see other humans, particularly those defeated in battle, as a valuable meat resource that can be exploited. This situation may arise in environments where animal protein is already a scarce commodity and conflict is cyclical and self-perpetuating, resulting in a relatively stable supply of meat. If these factors are coupled with the social sanctioning of necrophilia and anthropophagic activity (Spenneman 1990: 101) cannibalism may not be considered taboo.

Cultural connotations aside, the identification of cannibalism poses problems on a number of levels. As with recognition of conflict, it is imperative to distinguish cut marks arising from cannibalistic behaviour from other taphonomic agents. However, evidence for defleshing and subsequent consumption need not be restricted to cut mark analysis. Variegation in the pattern of burnt human bone has been used to speculate that flesh had been removed prior to burning (Baxter 1996).

The preceding discussion has served as a brief prologue to three significant research areas that cut mark analysis can contribute to, giving an indication of the depth of interpretation that this unique tool can lead to. What must be remembered is that, in general, all three of these components potentially interlink and show interdependence when humans disarticulate other humans. Accordingly, they should form an important part of the interpretative process. The ensuing sections will look at some of the factors that are important in arriving at the stage where patterns of activity can be inferred, and what is needed in order to validate these conclusions.

Chaîne opératoire of butchery – towards improved interpretation

A fundamental difference exists between the analysis of cut marks from faunal assemblages and those where there is evidence of cannibalism. Within zooarchaeology it is rarely, if ever, the case that the species comprising the assemblage is also the one responsible for creating it; an obvious dynamic in anthropophagic circumstances (White 1992: 102). This crucial aspects needs to be kept in mind as the already intricate task of determining how and why cut marks occur on bone is made considerably more complex with the addition of the conceptual attitudes of humans.

The two most salient aspects in determining anthropophagic activity are the observed perimortem indications and the assemblage's element representation (White 1992: 100). Closely linked to this is correctly distinguishing different categories of marks, for example slicing marks from chopping marks, and the accurate assessment of what implements have been used to make the marks.

A number of factors regarding more fundamental characteristics of cut mark analysis are worth mentioning before looking at issues of identification and implement usage. One of the most important issues when looking at butchery is that the marks are invariably incidental; the practitioner has not actively set out to leave a mark on the bone. There may be occasions where mutilation, ritual or mortuary activity involves deliberate incisions; however, these should be placed in a separate category as they are indicative of an entirely different pattern of activity. Cut marks following processing, be that part of a ritual,

mortuary, or (as is being focussed on here) subsistence practice, are not intentionally created.[2]

A number of issues are involved; firstly whether discussing processing in an ethnographic or modern butchery context, the skill of the practitioner is consistently reported as being of a high standard. The working knowledge of both soft and hard skeletal morphology means that it rarely takes a trained butcher (modern context) or hunter-gatherer more then a matter of seconds to locate the exact joint position, or the best place to make an incision in order to disarticulate a joint (Spennerman 1990: 116; Rival 1996: 148-151; Food and Agriculture Organisation Pub. 91 1991). As mentioned previously, due to overall morphology, it is likely that disarticulation of humans will follow, at least in parts, the procedure by which all mammals are processed. Furthermore, it is essential to recognise the importance of tools; these are often highly specialised and can involve complex manufacturing techniques. Of paramount importance is the prevention of damage to the cutting edge of any implement; this allows for a longer period of use, reduces the need for sharpen the blade and facilitates cutting of the meat. To maintain a sharp cutting edge, a 'butcher' will actively avoid bone as this will invariably dull the edge.

What must be remembered is that the specialist morphological knowledge, coupled with the skills of the person carrying out the disarticulation and the desire *not* to damage the blade (by avoiding contact with bone) all work towards an under-representation of cut marks. Therefore, anyone interested in looking at cut marks, also needs to be prepared to look at overall patterns of fragmentation (Knüsel and Outram 2004) and processing activity that does not leave tangible evidence.

Accurate Identification

Identification is a crucial aspect of cut mark analysis and involves not only the initial recognition of butchery, often very much dependent on the skill of the osteoarchaeologist, but also the division into distinct categories of cut mark type.

Accuracy of initial identification forms the first stage of interpreting cut marks. A V-shaped cross-section, for example, has long been held as a reliable marker for identification of marks made with a metal knife (Stewart 1979: 33). While problems exist with regard to distinguishing cut marks from other taphonomic factors, research has shown a high level of accuracy in recognising 'cut marks' *per se* from carnivore activity and percussion marks (Blumenschine *et al* 1996). These findings show that, with a relatively small amount of practice, not only can the distinction be made from other taphonomic indicators, but different aspects of processing can also be differentiated.

---

[2] Although mortuary rituals themselves may include aspects of cannibalism, the distinction being made here relates to the reasons behind the cut marks, and ultimately why the marks are present.

Other authors have outlined the main points for accurate identification of cut marks (White 1992: 100-163; Greenfield 1999; 2000; Lyman 2001: 294-352) in greater detail than space here permits. The following evaluation is restricted to aspects of identification that are important for understanding some of the broader perspectives that will aid with interpretation of the cut marks.

In order to make the most of cut mark analysis, one needs to think about the *process* of butchery from the initial stages of identification. This is especially relevant where humans are concerned as the factors involved are likely to be considerably augmented. What needs to be maintained is an appreciation for what the marks represent; potentially, a mark may be indicative of cause of death or marrow fracturing – or any of the stages between these two ends of the scale. It is this 'scale of activity' that needs to be appraised. It is not enough to record how many marks are evident, their depth and size, and where they occur; this information needs to be used as a basis for understanding, for example, factors such as what orientation the body was in during the disarticulation process.

Achieving this level of information involves incorporating a more in depth look at the marks, as well as a more holistic approach to interpretation. The striation pattern left by the implement (Greenfield 1999) needs to be analysed in order to accurately assess the direction of the cut and the orientation of the practitioner, body and implement. All too often, the likelihood that the body (or parts of the body) are subject to movement and constantly re-orientated during the disarticulation process goes unrealised. Reports are presented that have clearly failed to recognise that the body or carcass is not in 'normal anatomical position' during processing.

This leads to the next point, namely the need on the part of the osteoarchaeologist to have a viable working knowledge of the soft skeletal morphology of humans (or animals). One cannot hope to make an accurate appraisal of how and why a cut was made without knowing what muscular parts of the body were being removed, cut into or disarticulated. Having a 'viable' knowledge does not imply necessarily having the depth or the same type of knowledge about skeletal morphology that a meat processor has. The experience of the 'butcher'[3] is dedicated to one specific task, the most facile means of disarticulating and removing flesh from a carcass. This is a unique knowledge base and quite different to understanding the mechanisms of the body from an academic or even medical standpoint.

These two points are closely linked and should be used in tandem. Firstly, when butchery marks are noted on a bone, it should be orientate into correct anatomical position. The muscles, tendons and other soft tissues that

would have been present need to be taken into account and effectively 'overlaid' onto the bone. Only then can the analyst begin the process of evaluating what the mark might represent. At this stage it is possible to start understanding some of the finer aspects of the dismemberment process, for example what position the carcass was in during disarticulation. By starting with the bone in [approximate] anatomical position it is possible to see what direction the mark came from and whether there would have been a deviation from 'normal positioning' i.e. with the person lying on the ground, placed on an elevated platform, or even suspended. To illustrate this, taking a zooarchaeological example, chop marks to the inferior surface in the region of the neck / tubercle of the ribs, for example of cattle, are likely to indicate that the animal was suspended. It would be difficult to deliver a 'chopping cut' (using a cleaver) into this region unless the carcass was in a suspended position. It is probable that the practitioner was facing the ventral aspect of the carcass (evisceration would likely take place in this position), chopping from the caudal vertebrae and pelvis to the cranium in order to split the carcass. This type of information can also be useful in estimating how many individuals were involved in the disarticulation process. In this case, once the animal has been suspended only one person is needed to split the carcass. In a situation where a human was being disarticulated at ground level, it would be more difficult to make this estimation. Subtle differences in the way the cutting implements were used, or if the 'handedness' of the practitioners could be established, might give clues to how many individuals were involved. However, this would be a subjective approximation at best.

It is imperative that, where possible, the osteoarchaeologist factors in the likely cutting implements found within the geographic region or chronological period. This will have a considerable impact on the probable cuts found and it cannot be stressed enough how important an accurate appraisal of implements is to understanding cut marks.

Improving Interpretation

All too often the outcome in zooarchaeology, when looking at cut mark analysis, has been to simply note where and how many marks are present from any given assemblage. Human osteology on the other hand has invariably been far more adept at looking for the broader interpretive frameworks that might lead to a better understanding of what was actually taking place during the disarticulation process and what was causal to it. This approach has its problems if factors such as correct identification or accuracy in assigning implements are not clarified and can result in incorrect interpretation. However, there is the acceptance that complex dynamics are involved and that interpretation of at least some of these factors is possible.

It is important that the marks are described in detail; the information thus gleaned should then be used to evaluate factors such as:

---

[3] *Caveat*: this term is used loosely to describe a person versed in processing either human or animal bodies – it will invariably be the case that the skills associated with disarticulation will have been developed on animal carcasses, hence the use of the term butcher.

- How many different types of tools are evident?
- How many people might have been needed for the disarticulation process, and can this be estimated from the evidence?
- What does the grouping and location of the marks indicate about the type of activity being observed?
- Is it possible to detect whether the body was processed on the ground, on an elevated platform, or indeed suspended either from the neck or feet?
- Can other taphonomic indicators, such as evidence for burning or levels of fragmentation contribute to our understanding of the overall sphere of activity?
- Can a link be established between the methods of disarticulation used for humans with those used other animals within the same context / region / period?

The issues and queries outlined above barely touch the surface of what cut mark analysis potentially offers. Analysts must be prepared to pose the right questions and evaluate each criterion using rigorous methods. The following ethnographic example offers a pertinent illustration of some of the information potentially available.

Spennerman (1990) recorded and analysed an assemblage of bones from Fiji, where both historical and ethnographic information have indicated that human flesh was consumed, particularly following conflict. The cut marks noted from the assemblage were very faint and would have gone unnoticed had it not been for the particular provenance of the assemblage. Due to the level of preservation, Spennerman was able to rule out pathological and other taphonomic factors as the cause for the modifications. All marks recorded were at, or near, joint articulations. This information was also viewed in light of the culinary practice of the indigenous peoples, who invariably cooked in earth ovens.

From this information Spennerman was able to decipher that the implements used were made from bamboo, an unusual material from a western perspective, but one that has been widely used in South East Asia and parts of Oceania. The blade is sharpened by removing a sliver of bamboo to leave a new cutting edge; however, due to the nature of the material it tends to leave very faint marks on the bone. The method of cooking did not require the meat to be processed beyond 'gross disarticulation' as large joints, and even whole pigs or humans, were known to be cooked in this way (Spennerman 1990: 108-134).

The above study demonstrates how a combination of accurate recognition coupled with aspects of tool manufacture and cooking practice, allowed for a convincing reconstruction of the overall process of disarticulation. The nature of the implements, along with the specific type of food preparation, formed the catalysts for the pattern of butchery seen. The implements could not cut bone in the same way that metal or flint tools do; nor was there any need for de-fleshing cuts as the cooking methods called for large joints of meat. Marks for meat removal, common in European contexts for example, were not evident in this type of assemblage as the meat, once cooked, was easily removed from the bone without the need for further cutting. What this project demonstrates is that factors such as provenance, implements and cooking practice are all essential criteria and need to be brought into the framework of interpretation. Only then will an accurate appraisal of the cut marks be established.

Conclusion

In conclusion, this article has provided an evaluation of the main areas where cut mark research can make a useful contribution to the investigation of human remains. Unfortunately, there are few specialists who focus specifically on the attributes of processing marks or on their potential for further interpretation; although it is fair to say that the majority of osteoarchaeologists do recognise the value of studying cut marks.

It should once again be reiterated that a good knowledge of skeletal soft tissue anatomy is essential when looking at butchery marks, as is an understanding of what implements were likely to be present and how they were used. Anyone interested in developing this line of research must gain a clearer understanding and appreciation for how important 'process' is to cut mark analysis. Furthermore, one must also be prepared to develop a knowledge of implements; specifically the implements likely to have been present and in use in the region/site/period under investigation.

It has been demonstrated that this line of research has broad application and is not restricted to evaluations of subsistence. It can certainly contribute to interpretations of injury and violence, which although rarely discussed in an anthropological context (Larsen 1997: 154), would gain a great deal by adopting a rigorous methodological approach to analysing the relevant cut marks.

Acknowledgements

I would like to acknowledge the advice and generous support of my supervisor, Preston Miracle, for all his assistance throughout my on-going doctoral research. The guidance of my former Masters Supervisor, Mark Maltby who is gratefully accredited for directing my past experience as a butcher into this specific research area. I would like to thank Louise Loe not only for inviting me to present this research but also for reading this article and offering many insightful comments. I am grateful to Aleks Pluskowski, and Joanne Bennett for their time and patience in reading through this article and providing much needed lucidity. Thanks also to Alice Roberts and Kate Robson-Brown for this generous opportunity, and to my college, Peterhouse, for their support and aid.

References

Andrews P & Fernandez-Jalvo Y. 1997. Surface modifications of the Sima de los Huesos fossil humans. *Journal of Human Evolution*: 33; 191-217.

Baxter IL. 1996. Evidence of ritual and magic in Leicestershire & Rutland during the prehistoric period. In *Ritual Treatment of Humans and Animal Remains.*, Anderson S and Boyle K (eds.). Proceedings of the 1st Meeting of the Osteoarchaeological Research Group. Oxbow: Oxford.

Binford LR. 1981. *Bones: Ancient Men and Modern Myths*. Academic Press: New York.

Boylston A. 2000. Evidence for weapon-related trauma in British archaeological samples. In *Human Osteology in Archaeological and Forensic Science.*, Cox M and Mays S (eds.). Greenwich Medical Media: London; 357-381.

Blumenschine RJ, Marean CW, Capaldo SD. 1996. Blind tests of inter-analyst correspondence and accuracy in identification of cut marks, percussion marks and carnivore tooth marks on bone surfaces. *Journal of Archaeological Science*: 23; 493-507.

Food and Agriculture Organization. 1991. *Guidelines for Slaughter, Meat Cutting and Further Processing*. Publication 91. Rome.

Greenfield HJ. 1999. The origins of metallurgy: distinguishing stone from metal cut-marks on bones from archaeological sites. *Journal of Archaeological Science*: 26; 797-808.

Hurlbut SA. 2000. The taphonomy of cannibalism: A review of anthropogenic bone modification in the American southwest. *International Journal of Osteoarchaeology*: 10; 4-26.

Howell S. 1996. Nature in culture or culture in nature? Chewong ideas of 'humans' and other species. In *Nature and Society: Anthropological Perspectives.*, Descola P and Palsson G (eds.). Routledge: London; 127-144.

Ingold T. (ed). 1988. *What is an Animal?* Unwin Hyman: London.

Kennedy KAR. 1994. Identification of sacrificial and massacre victims in archaeological sites: the skeletal evidence. *Man and Environment*: 19: 247-51.

Knüsel CJ & Outram AK. 2004. Fragmentation: the zonation method applied to fragmented human remains from archaeological and forensic contexts. *Environmental Archaeology*: 9:1; 83-97.

Larsen CS. 1997. *Bioarchaeology: Interpreting Behaviour from the Human Skeleton*. Cambridge University Press: Cambridge.

Lyman RL. 1978. Prehistoric butchering techniques in the lower Granite Reservoir, south eastern Washington. *Tebiwa*: 13; 1-25.

Lyman RL. 1987. Archaeofaunas and butchery studies: a taphonomic perspective. In *Advances in Archaeological Methods and Theory*, Vol 10. Academic Press. New York.

Lyman RL. 2001. *Vertebrate Taphonomy*. Cambridge University Press: Cambridge.

Ogilvie MD & Hilton CE. 2000. Ritualized violence in the prehistoric American Southwest. *International Journal of Osteoarchaeology*: 10; 27-46.

Pluskowski A. (ed). 2005. *Just Skin and Bones? New Perspectives on Human-Animal Relations in the Historic Past*. BAR: Oxford.

Rabett RJ & Piper PJ. n.d. Eating your tools: Early butchery and craft modification of primate bones in tropical Southeast Asia. In *Bones for Tools, Tools for bones: Hunter-Gatherer Resource Exploitation.*, Seetah K and Gravina BK (eds.). In prep.

Rival L. 1996. Blowpipes and Spears: The social significance of Huaorani technological choices. In *Nature and Society: Anthropological Perspectives.*, Descola P and Palsson G (eds.). Routledge: London; 145-164.

Seetah K. 2005. Butchery as a tool for understanding the changing views of animals. In *Just Skin and Bones? New Perspectives on Human-Animal Relations in the Historic Past.*, Pluskowski. A. (ed). BAR: Oxford.

Spennemann DHR. 1990. Don't forget the bamboo. On recognising and interpreting butchery marks in tropical faunal assemblages, some comments asking for caution. In *Problem Solving in Taphonomy: Archaeological and Palaeontological Studies from Europe, Africa and Oceania.*, Solomon S., Davidson I and Watson D. (eds.). Tempus: Archaeological and Material Cultural Studies in Anthropology 2:108-134.

Stewart TD. 1979. *Essentials of Forensic Anthropology*. Charles C. Thomas. Springfield.

Thompson TJU. 2002. The assessment of sex in cremated individuals: Some cautionary notes. *Canadian Society of Forensic Science Journal*: 35(2); 49-56.

Thompson TJU. 2004. Recent advances in the study of burned bone and their implications for Forensic Anthropology. *Forensic Science International*: 146S; S203–S205.

Wells C. 1964. *Bones, Bodies and Disease*. Thames and Hudson: London.

White TD. 1992. *Prehistoric Cannibalism at Mancos 5MTUMR-2346*. Princeton University Press: New Jersey.

www.ingramcontent.com/pod-product-compliance
Lightning Source LLC
Chambersburg PA
CBHW061302270326
41932CB00029B/3445